CalcLabs with Mathematica

for

Stewart's

Multivariable Calculus

Sixth Edition

Selwyn Hollis
Armstrong Atlantic State University

THOMSON

BROOKS/COLE

Australia • Brazil • Canada • Mexico • Singapore • Spain • United Kingdom • United States

Printed in the United States of America

1 2 3 4 5 6 7 11 10 09 08 07

Printer: Thomson/West
Cover Image: © M. Neugebauer/zefa/Corbis

ISBN-13: 978-0-495-11890-9
ISBN-10: 0-495-11890-7

Thomson Higher Education
10 Davis Drive
Belmont, CA 94002-3098
USA

For more information about our products, contact us at:
Thomson Learning Academic Resource Center
1-800-423-0563

For permission to use material from this text or product, submit a request online at
http://www.thomsonrights.com.
Any additional questions about permissions can be submitted by email to **thomsonrights@thomson.com.**

Contents

4 Multivariate Functions   73

5 Multiple Integrals   113

6 Vector Calculus   142

Prologue

This is a manual written to accompany the sixth edition of James Stewart's *Calculus*, one of the most respected and successful calculus texts of recent years—a time during which the wide availability of powerful computational software such as *Mathematica*® has had a significant effect upon the way calculus is taught and learned. Remarkably, computations that would have been extraordinary just a few years ago—and certainly unimaginable in the times of Newton, Gauss, and Riemann—can be done easily by any student with a laptop computer and *Mathematica*.

The primary goal of this manual is to show you how *Mathematica* can help you learn and use calculus. The approach (we hope) is not to use *Mathematica* as a "black box," but rather as a tool for exploring calculus concepts and the way calculus can be used to solve problems.

Two secondary goals of this manual are: 1) to present in a very concise manner the central ideas of calculus, and 2) to introduce you to many of the capabilities of *Mathematica*. You should be aware of—but hopefully not intimidated by—the fact that *Mathematica* is an enormously complex system that can be frustrating to the beginner. However, with discipline and a little perseverance, you will soon begin to see the basic elegance and (believe it or not) underlying simplicity of *Mathematica*.

The last chapter of this manual contains twenty-four extended exercises, or "projects," which cover a wide range of topics and are arranged roughly in the same order as the corresponding material in Stewart's *Calculus*. The projects vary considerably in length, level of difficulty, and the amount of guidance provided. We hope that these projects will be interesting—and sometimes fun—while reinforcing important calculus concepts.

Mathematica will not "do calculus" for you. It cannot decide the proper approach to a problem, nor can it interpret results for you. In short, *you* still have to do the thinking, and *you* need to know the fundamental concepts of calculus. You must learn calculus from lectures and your textbook—and most importantly by working problems. That's where *Mathematica* comes in as a learning tool, allowing you to concentrate on concepts and to work interesting problems without getting bogged down in algebraic and computational details.

Until you've had a lot of experience using *Mathematica*, you will need to consult the *Mathematica* Documentation Center frequently. The first thing you should do is familiarize yourself with the Documentation Center, in which you can find detailed descriptions of essentially every element of *Mathematica*, as well as a wealth of tutorial material. Also, the appendix at the end of this manual is an introduction to essentially all of the basic *Mathematica* concepts necessary for working your way through the manual.

■ Web Site

Corrections to this manual, as well as related updates, *Mathematica* notebooks, and add-ons will be posted at www.math.armstrong.edu/faculty/hollis/mmacalc/ . This is where you can find the `Vectors` package that is used extensively in this manual.

■ *Mathematica 6*

This, the third edition of this manual, was produced entirely with *Mathematica* 6 (6.0.0) and relies upon many of its new features. Hence, any attempt to use this manual with an earlier version of *Mathematica* will only lead to confusion and frustration.

Mathematica 6 is a revolutionary new version of *Mathematica*. The sheer number of new features and functions is astonishing. Even *Mathematica* users with years of experience have a vast new world to explore. On the other hand, many of the changes have made using *Mathematica* much simpler and more intuitive. In fact, most commands in this manual have to some degree been shortened, and many significantly so.

The following briefly describes (for those with prior *Mathematica* experience) some of the new ways of doing things that are particularly pertinent to this manual.

■ Displaying and Combining Graphics

The process of combining graphics has been greatly simplified, since the display of graphics is no longer a "side effect" of the creation of a graphics object. The following describe a few of the most basic consequences.

- Graphics may be suppressed by a semicolon just as any other output. For instance, the following produces no picture or other output.

```
Plot[Sin[x], {x, 0, 2 π}];
```

- Graphics computed inside of Show are not displayed individually prior to the output of Show.

```
Show[Plot[x², {x, -1, 1}],
  Graphics[Circle[{0, .5}, .5]], AspectRatio → Automatic]
```

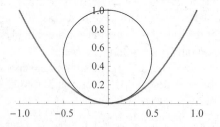

Hence there is no longer any need for the DisplayTogether function or the DisplayFunction option.

- The function Graphics (and Graphics3D) produces visual output without the help of Show.

```
Graphics[{Gray, Disk[]}, ImageSize → 48]
```

■ The Standard Packages

Most of the functions that were previously available in the standard packages either have been incorporated into the *Mathematica* kernel or their functionality has been assumed by other functions. The following are a couple of pertinent examples.

- The `Graphics`FilledPlot`` package is gone. `Plot` now has a `Filling` option.

```
Plot[{Sin[x], Sin[2 x], Sin[3 x]},
  {x, 0, 2 π}, Filling → {1 → {3}, 1 → {2}}]
```

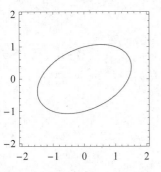

- The `Graphics`ImplicitPlot`` package is gone. `ContourPlot` now works on equations.

$$\texttt{ContourPlot}\left[x^2 - x\,y + 2\,y^2 == 2,\ \{x,\ -2,\ 2\},\ \{y,\ -2,\ 2\}\right]$$

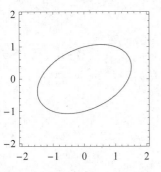

■ Animate and Manipulate

Making animations is now remarkably simple. `Animate` (which was previously in the `Graphics`Animation`` package) eliminates the need to plot individual frames and collapse the resulting celll group. `Animate` works much like `Table`. (Note that `Table` now *literally* produces a table of graphics.)

$$\texttt{Table}\left[\texttt{Plot}\left[x^n,\ \{x,\ 0,\ 1\},\ \texttt{PlotRange} \to \{0,\ 1\}\right],\ \{n,\ 1,\ 4\}\right]$$

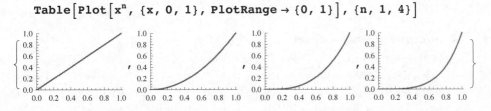

Here's the same command with `Table` replaced by `Animate`.

$$\texttt{Animate}\left[\texttt{Plot}\left[\texttt{x}^{\texttt{n}}\texttt{, \{x, 0, 1\}, PlotRange} \rightarrow \texttt{\{0, 1\}}\right]\texttt{, \{n, 1, 4\}}\right]$$

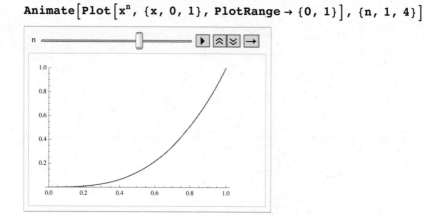

While the default increment for `Table` is 1, `Animate` changes *continuously* by default.

A similar new function is `Manipulate`, which provides a slider (or sliders) for the user to "manually" change one or more parameter values. It is also possible to specify different types of controls such as radio buttons and pop-up menus, as well as "locators," with which one can manipulate positions of points within the graphic itself. For instance, the following shows the graph of the parabola through three points and allows live manipulation of all three points.

```
Manipulate[
 Plot[InterpolatingPolynomial[{pt1, pt2, pt3}, x],
  {x, -1, 1}, PlotRange → {{-1, 1}, {-1, 1}}],
 {{pt1, {-.5, .5}}, Locator}, {{pt2, {0, -.5}}, Locator},
 {{pt3, {.5, .5}}, Locator}]
```

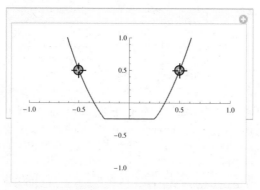

■ The Documentation Center

The old Help Browser has been replaced by the *Documentation Center*, which has greater search capability and the ability to be automatically updated via the internet. The Documentation Center contains a lot of helpful tutorials, but at the time of this publication how to find those tutorials is less than obvious. We hope that direct links to the tutorials will be added to the "front page" of the Documentation Center in subsequent updates of *Mathematica* 6.

1 Vectors

1.1 Visualizing Vectors with *Mathematica*

Vectors are represented in *Mathematica* by lists. A two-dimensional vector is a list of length two and a three-dimensional vector is a list of length three. The following is the *Mathematica* definition of a particular two-dimensional vector $\mathbf{a} = \langle a_1, \, a_2 \rangle$.

```
a = {3, 1}
```

```
{3, 1}
```

The zero vector (or the point (0, 0)) is defined by

```
0⃗ = {0, 0}
```

```
{0, 0}
```

In *Mathematica* there is no distinction between a vector and a point; each is represented by a list. So if we define a = {3,1} as above, that list could represent either the point $A(3, 1)$ or the position vector $\langle 3, 1 \rangle$ of that point.

Mathematica's `Arrow` primitive allows us to plot representations of two-dimensional vectors by specifying an initial point and a terminal point. This plots **a** as a position vector:

```
Show[Graphics[Arrow[{0⃗, a}]], Axes → True,
  PlotRange → {{-1.5, 3.5}, {-.5, 1.5}}]
```

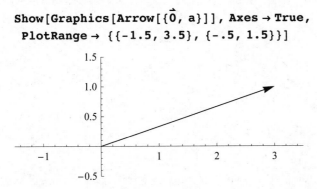

We can simplify the plotting of vectors by defining a function called `Vector` that will take a vector and return a corresponding graphics object. The vector will be specified as a pair of ordered pairs, with each ordered pair being of the form {*initial point, vector*}. The definition will have two parts. This defines `Vector` with optional style directives:

```
Vector[{ip : {_, _}, vec : {_, _}}, styles___] := Graphics[
  {Arrowheads[Medium], Thickness[.005], styles,
  Arrow[{ip, ip + vec}] } ]
```

Finally, this makes specification of the initial point unnecessary for position vectors:

```
Vector[posvec : {_, _}, styles___] := Vector[{{0, 0}, posvec}, styles]
```

Now let's plot **a** as a position vector along with the representation of **a** with initial point $P(-1, 1)$.

```
p = {-1, 1}; a = {3, 1}; Show[Vector[a, Blue], Vector[{p, a}, Red],
    Axes → True, PlotRange → {{-1.5, 3.5}, {-.5, 2}}]
```

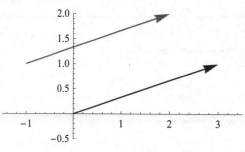

Mathematica's `Table` command provides a convenient way of plotting a family of vectors. Here's an interesting picture that shows the position vectors of ten points along the parabola $y = x^2$.

```
Show[Table[Vector[{.1 i, (.1 i)²}, Hue[.1 i]], {i, 1, 10}],
    Axes → True]
```

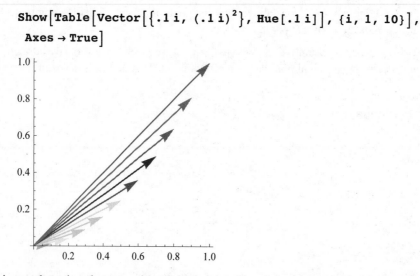

It is worth noting that any given initial point $P(x_1, y_1)$ and terminal point $Q(x_2, y_2)$ determine a vector

$$\mathbf{a} = \langle x_2 - x_1, \ y_2 - y_1 \rangle.$$

The following is a plot of two representations of the vector determined by the initial point $P(2, -2)$ and the terminal point $Q(9, 1)$. One has the initial point P, and the other is the position vector of the point $A\,(7, 3)$.

```
p = {2, -2}; q = {9, 1}; a = q - p;
Show[Vector[{p, a}], Vector[a], Axes → True]
```

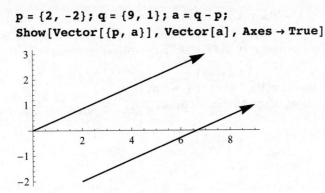

■ Three-Dimensional Vectors

To plot three-dimensional vectors, we'll need the **Vectors** package. (*Important note*: This is not one of the standard *Mathematica* packages. See the Prologue for information.)

```
<< Vectors`
```

This package extends the previous definition of **Vector** to plot three-dimensional vectors. (It also includes the previous 2-D definition.) Here's the position vector of the point (3, 1, 1).

```
Vector[{3, 1, 1}]
```

This plots vectors **i**, **j**, and **k** as position vectors.

```
i = {1, 0, 0}; j = {0, 1, 0}; k = {0, 0, 1};

Show[Vector[i], Vector[j], Vector[k], ViewPoint → {6, 2, 2}]
```

Note that setting **ViewPoint→{6,2,2}** provides a view that is consistent with the usual one in calculus texts, in which the *x*-axis (or the **i** vector) points "out" of front side the page, slightly down and to the left.

For convenience the `Vectors` package defines `ijk` to be the result of applying `Vector` to **i**, **j**, and **k**. Here's a plot of **a** = $\langle 1, 1, 0 \rangle$ and **b** = $\langle -1/2, -1/2, 1 \rangle$ as a position vectors together with **b** located at **a**, **a** located at **b**, and **i**, **j**, and **k** as position vectors.

```
a = {1, 1, 0}; b = {-1/2, -1/2, 1};
Show[Vector[a], Vector[b], Vector[{a, b}],
  Vector[{b, a}], ijk, ViewPoint → {6, 2, 2}]
```

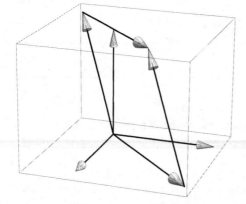

1.2 Length and Arithmetic Operations

Discussion of the notions in this section is found in Section 13.2 of Stewart's *Calculus*. (ET:12.2)

The **length** (or *norm*) of a two-dimensional vector **a** = $\langle a_1, a_2 \rangle$ is

$$\| \mathbf{a} \| = \sqrt{a_1^2 + a_2^2} \ .$$

Similarly, the length of a three-dimensional vector **a** = $\langle a_1, a_2, a_3 \rangle$ is

$$\| \mathbf{a} \| = \sqrt{a_1^2 + a_2^2 + a_3^2} \ .$$

Mathematica's `Length` function does not compute the length of a vector; instead it gives the number of items in a list. The length function for vectors is `Norm`. So, for example, the length of the vector $\langle 4, -3, 1 \rangle$ is

```
Norm[{4, -3, 1}]
```

$\sqrt{26}$

Addition of vectors is handled automatically by *Mathematica*. For example,

```
a = {2, 3}; b = {2, -1}; a + b
```

```
{4, 2}
```

The following plots **a**, **b**, and **a + b** as position vectors. (Notice how we *map* Vector through the list of vectors with /@.)

`Show[Vector /@ {a, b, a + b}, Axes → True]`

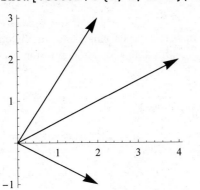

Notice that **a + b** is a diagonal of the parallelogram formed by **a** and **b**. In fact, this becomes even clearer if we plot additional representations of **a** and **b**.

`Show[Vector /@ {a, b, a + b, {b, a}, {a, b}}, Axes → True]`

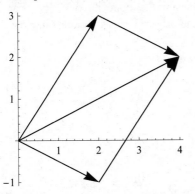

The following is an analogous picture involving three-dimensional vectors.

```
a = {-1, 2, 1 / 2}; b = {1, -1, 1};
Show[ijk, Vector /@ {a, b, a + b, {b, a}, {a, b}},
  ViewPoint → {6, 2, 2}]
```

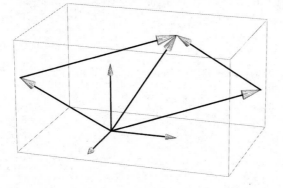

Multiplication of a vector by a scalar is handled automatically by *Mathematica*; for example,

```
a = {1, 2, 3}; 3 a
{3, 6, 9}
```

Multiplying a vector by a positive scalar can only change its length, not its direction. For example, if we plot the vector $\mathbf{a} = \langle 2, 1 \rangle$, along with $2\mathbf{a}$ and $3\mathbf{a}$, all as position vectors, we see the following.

```
a = {2, 1};
Show[Vector /@ {a, 2 a, 3 a}, Axes → True]
```

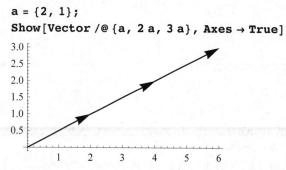

Multiplying a vector by a negative scalar gives it the opposite direction. For position vectors, the effect is that of reflection through the origin. Here is a plot of $\mathbf{a} = \langle 2, 1 \rangle$ along with $-\mathbf{a}$.

```
Show[Vector /@ {a, -a}, Axes → True]
```

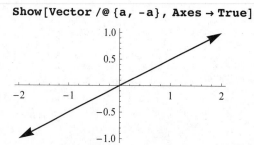

Subtraction of vectors. Let's plot two vectors $\mathbf{a} = \langle 4, 1 \rangle$ and $\mathbf{b} = \langle 2, 2 \rangle$ along with their difference $\mathbf{a} - \mathbf{b}$. Since $\mathbf{a} = (\mathbf{a} - \mathbf{b}) + \mathbf{b}$, it is not surprising to see that $\mathbf{a}$ is a diagonal of the parallelogram formed by $\mathbf{a} - \mathbf{b}$ and $\mathbf{b}$.

```
a = {4, 1}; b = {2, 2};
Show[Vector /@ {a, b, a - b}, Axes → True]
```

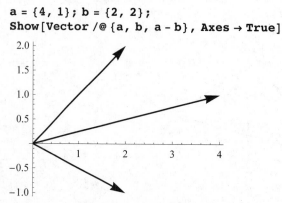

We are more interested, however, in the geometric relationship of **a** − **b** to **a** and **b**. So notice that if we translate **a** − **b** so that its initial point is at the terminal point of **b**, then the terminal point of **a** − **b** is precisely the terminal point of **a**.

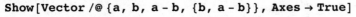

```
Show[Vector /@ {a, b, a - b, {b, a - b}}, Axes → True]
```

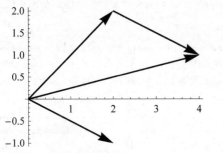

Therefore, **a** − **b** is just the *other diagonal* of the parallelogram formed by **a** and **b**. Here's a plot of **a**, **b**, and **a** − **b** as position vectors, and each of the vectors **a** + **b** and **a** − **b** as a diagonal of the parallelogram formed by **a** and **b**.

```
Show[Vector /@ {a, b, a - b, {b, a - b}, a + b, {a, b}, {b, a}},
  Axes → True]
```

This is a similar picture of three-dimensional vectors.

```
a = {1, 1, 0}; b = {0, 1, 1};
Show[Vector /@ {a, b, a - b, {b, a - b}, a + b, {a, b}, {b, a}},
  ijk, ViewPoint → {6, 2, 2}]
```

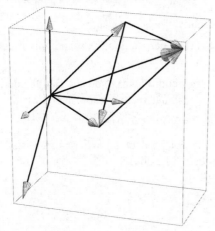

1.3 The Dot Product and Projections

Detailed discussion and definitions of the notions in this section are found in Section 13.3 of Stewart's *Calculus*. (ET: 12.3)

Given two nonzero vectors **a** and **b** (either two- or three-dimensional), when we speak of the angle between them, we mean the angle θ, $0 \leq \theta \leq \pi$, formed at the origin by their representations as position vectors. For example, basic trigonometry tells us that the angle between **a** = $\langle \sqrt{3}, 1 \rangle$ and **b** = $\langle -\sqrt{3}, 1 \rangle$ is $\theta = 2\pi/3$.

```
a = {√3 , 1}; b = {-√3 , 1};
Show[Vector /@ {a, b}, Axes → True,
  PlotRange → {{-2, 2}, {-.25, 1.25}}, Ticks → {{-1, 1}, {1}}]
```

The **dot product** of two vectors **a** and **b** is defined by

$$\mathbf{a} \cdot \mathbf{b} = \|\mathbf{a}\| \, \|\mathbf{b}\| \, \cos \theta,$$

where θ is the angle between **a** and **b**. If either of **a** or **b** is zero, then θ is undefined; so we define $\mathbf{a} \cdot \mathbf{b} = 0$ in that case.

For the example above in which **a** = $\langle \sqrt{3}, 1 \rangle$ and **b** = $\langle -\sqrt{3}, 1 \rangle$, each vector has length 2 and $\cos \theta = -1/2$; so $\mathbf{a} \cdot \mathbf{b} = (2)(2)(-1/2) = -2$.

Note that two interesting facts follow immediately from this definition. First, when **a** = **b**, we have $\cos \theta = 1$ (since $\theta = 0$), and so

$$\mathbf{a} \cdot \mathbf{a} = \|\mathbf{a}\|^2, \text{ or equivalently, } \|\mathbf{a}\| = \sqrt{\mathbf{a} \cdot \mathbf{a}}.$$

Second, non-zero vectors **a** and **b** are orthogonal (i.e., perpendicular), if and only if $\cos \theta = 0$ (since $\theta = \pi/2$), which happens precisely when $\mathbf{a} \cdot \mathbf{b} = 0$.

There is a much simpler formula in terms of vector coordinates. You can find the derivation in Section 9.3 of Stewart. For two- and three-dimensional vectors respectively, we have

$$\mathbf{a} \cdot \mathbf{b} = a_1 b_1 + a_2 b_2 \text{ and } \mathbf{a} \cdot \mathbf{b} = a_1 b_1 + a_2 b_2 + a_3 b_3.$$

The dot product is built into *Mathematica*. The dot product of two vectors (lists) can be computed by simply placing a period between them, like this:

```
Clear[a, b]; {a₁, a₂}.{b₁, b₂}
```

$a_1 b_1 + a_2 b_2$

Now, since we have this way of computing dot products, we can solve for θ in the definition of the dot product, obtaining a formula for the angle between two vectors:

$$\theta = \arccos\left(\frac{\mathbf{a}\cdot\mathbf{a}}{\|\mathbf{a}\|\,\|\mathbf{b}\|}\right) = \arccos\left(\frac{\mathbf{a}\cdot\mathbf{a}}{\sqrt{\mathbf{a}\cdot\mathbf{a}}\,\sqrt{\mathbf{b}\cdot\mathbf{b}}}\right)$$

■ Example 1.3.1

Find the angle between the vectors $\mathbf{a} = \langle 5, 1\rangle$ *and* $\mathbf{b} = \langle -1, 3\rangle$.

Let's first look at a picture of the situation by plotting **a** and **b** as position vectors.

```
a = {5, 1}; b = {-4, 3};
Show[Vector[a], Vector[b], Axes → True]
```

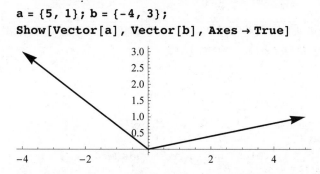

Now using the above formula, we'll compute the angle between **a** and **b** followed by numerical approximation and conversion to degree measure.

```
ArcCos[a.b / √((a.a) (b.b))]

N[%]
% / Degree
```

$$\mathrm{ArcCos}\left[-\frac{17}{5\sqrt{26}}\right]$$

```
2.3007
```

```
131.82
```

■ Projections

One of the most important uses of the dot product is the calculation of the projection of one vector onto another. Let **a** and **b** be two vectors (two- or three-dimensional). The **(vector) projection of b onto a** is the vector that has the same direction as **a** and together with **b** forms a right triangle. One example is illustrated in the following plot.

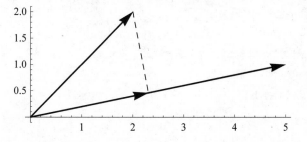

The **component of b along a** is calculated as follows:

$$\texttt{comp[b\_, a\_] := a.b} \Big/ \sqrt{\texttt{a.a}}$$

The (vector) projection of **b** onto **a** is then

$$\texttt{proj[b\_, a\_] := comp[b, a] a} \Big/ \sqrt{\texttt{a.a}}$$

Be careful to note that the component of **b** along **a** is a scalar, while the projection of **b** onto **a** is a vector. The component of **b** along **a** is in fact either the length of the projection of **b** onto **a** or its negative.

By definition, the triangle formed by **b** and the projection of **b** onto **a** is a right triangle:

```
proj[b, a].(b - proj[b, a]) // Simplify
```
0

The following is a command for plotting two vectors along with the projection of one onto the other.

```
showProj[b_, a_, opts___] :=
 Show[Vector[proj[b, a], Thick, Red, Arrowheads[Medium]],
  Vector[a], Vector[b],
  Graphics[{Dashed, Line[{b, proj[b, a]}]}], opts, Axes → True]
```

■ Example 1.3.2

For the vectors **a** and **b** in the preceding picture, find the component of **b** along **a** and the projection of **b** onto **a**. Then find the component of **a** along **b** and the projection of **a** onto **b**.

This computes the component of **b** along **a** and the projection of **b** onto **a**:

```
a = {5, 1}; b = {2, 2};
{comp[b, a], c = proj[b, a]}
```

$$\left\{ 6\sqrt{\frac{2}{13}},\ \left\{ \frac{30}{13},\ \frac{6}{13} \right\} \right\}$$

Here's the picture:

```
showProj[b, a]
```

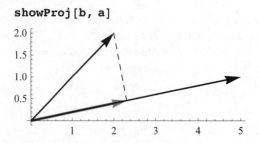

Here are the other computations:

```
{comp[a, b], c = proj[a, b]}
```

$$\left\{ 3\sqrt{2},\ \{3, 3\} \right\}$$

This is the other picture:

showProj[a, b]

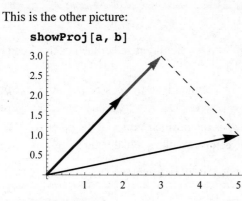

■ **Example 1.3.3**

For the three-dimensional vectors

$$\mathbf{a} = \langle 1, 2, 1 \rangle \text{ and } \mathbf{b} = \langle 1, 1, 1 \rangle,$$

find and plot the component of **b** *along* **a** *and the projection of* **b** *onto* **a**.

Here are the computations:

```
a = {1, 2, 1}; b = {1, 1, 1};
comp[b, a]
c = proj[b, a]
```

$$2\sqrt{\frac{2}{3}}$$

$$\left\{ \frac{2}{3}, \frac{4}{3}, \frac{2}{3} \right\}$$

A modification of the previous `showProj` command will create the picture.

```
showProj3D[b_, a_, opts___] := Show[ijk, Vector[a],
   Vector[b], Vector[proj[b, a], Red, Thick], Graphics3D[
     {Dashed, AbsoluteThickness[1], Line[{b, proj[b, a]}]}],
   opts, ViewPoint → {6, 2, 2}]
```

showProj3D[b, a]

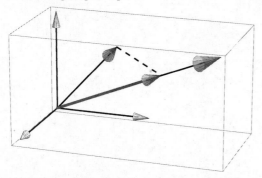

Note that in this example the component of **b** along **a** is negative, and so the projection of **b** onto **a** points in the direction opposite that of **a**.

◆ Exercises

1. Enter the (2D) definition(s) of `Vector` and test it by plotting together the position vectors $\langle 1, 1 \rangle, \langle 1, 1 \rangle, \langle -1, 1 \rangle, \langle 1, -1 \rangle$, and $\langle -1, -1 \rangle$.

2. Use `Vector` to make a plot showing position vectors given by $\langle \cos(k\pi/6), \sin(k\pi/6) \rangle$, $k = 1, \ldots, 12$.

3. Let a_k be as in Exercise 2 and suppose that a_k is the position vector of the point A_k. Use `Vector` to make a plot showing each of the vectors a_k with its initial point at A_k.

4. Use `Vector` to make a plot showing position vectors $\left\langle \frac{1}{k} \cos(k\pi/6), \frac{1}{k} \sin(k\pi/6) \right\rangle$, $k = 1, \ldots, 24$

5. Use `Vector` to make a plot showing position vector $\langle k - 5, k/2 \rangle$, $k = 1, \ldots, 10$.

6. Load the `Vector` package and plot together the position vectors $\langle 1, 1, 0 \rangle, \langle 1, 0, 1 \rangle$, and $\langle 0, 1, 1 \rangle$.

7. Use `Vector` to make a plot showing position vectors given by $\langle \cos(k\pi/6), \sin(k\pi/6), 1 \rangle$, $k = 1, \ldots, 12$.

8. Use `Vector` to make a plot showing position vectors given by $\langle t, 2 - t, 2t \rangle$, $t = -.5, -.25$, $0, \ldots, 1.25, 1.5$. Of all such vectors, where t is any real number, find the one with minimum length.

9. Let $\mathbf{a} = \langle 1, 2, 5 \rangle$ and $\mathbf{b} = \langle 1, -2, -3 \rangle$. Compute $\|\mathbf{a}\|, \|\mathbf{b}\|, \mathbf{a} \cdot \mathbf{b}$, the angle between $\mathbf{a}$ and $\mathbf{b}$, the projection of $\mathbf{b}$ along $\mathbf{a}$, and the projection of $\mathbf{a}$ onto $\mathbf{b}$.

10. Verify that the vectors $\langle t, \sin t, 1 - t \rangle$ and $\langle t - 1, (t - 1)\cos t, t + \cos t \sin t \rangle$ are orthogonal for all real numbers t.

11. Verify that $\langle -s + t, s + 2t, t - 3s \rangle$ is orthogonal to $\langle 7, -2, -3 \rangle$ for all real numbers s and t.

12. Find each of the two 2D unit vectors that form an angle of $\pi/6$ radians with $\langle 2, 3 \rangle$.

1.4 The Cross Product

Discussion of the cross product of two vectors is found in Section 13.4 of Stewart's *Calculus*. (ET:12.4)

Given two non-zero, non-parallel, three-dimensional vectors **a** and **b**, the **cross product** of **a** and **b** is the vector

$$\mathbf{a} \times \mathbf{b} = (\|\mathbf{a}\| \|\mathbf{b}\| \sin\theta)\, \mathbf{n},$$

where **n** is the unit vector that is perpendicular to both **a** and **b**, with direction given by the "right-hand rule." We define $\mathbf{a} \times \mathbf{b} = \mathbf{0}$ if either **a** or **b** is **0** or if **a** and **b** are parallel. (Note that $\sin\theta = 0$ if **a** and **b** are parallel.) By definition, a nonzero cross product $\mathbf{a} \times \mathbf{b}$ is perpendicular to both **a** and **b**, since it is a scalar multiple of **n**. *Mathematica* has a built-in function Cross for computing cross products; the result is in "component form:"

```
Clear[a, b]; Cross[{a₁, a₂, a₃}, {b₁, b₂, b₃}]
```
$$\{-a_3 b_2 + a_2 b_3,\ a_3 b_1 - a_1 b_3,\ -a_2 b_1 + a_1 b_2\}$$

Cross products can also be entered in StandardForm rather than InputForm. (The cross symbol can be entered by clicking on a button in either of the standard BasicInput or BasicTypesetting palettes, or by typing [ESC]cross[ESC].) Here are two examples:

```
{a₁, a₂, a₃} × {b₁, b₂, b₃}
```
$$\{-a_3 b_2 + a_2 b_3,\ a_3 b_1 - a_1 b_3,\ -a_2 b_1 + a_1 b_2\}$$

```
{1, 2, 3} × {3, 1, 5}
```
$$\{7, 4, -5\}$$

The following picture shows a pair of vectors **a** and **b**, and then **a** and **b** along with $\mathbf{a} \times \mathbf{b}$.

```
a = {2 / 3, 2 / 3, -1 / 4}; b = {0, 1, 1 / 2}; Needs["Vectors`"];
GraphicsRow[{Show[ijk, Vector /@ {a, b}, ViewPoint → {6, 2, 2}],
    Show[ijk, Vector /@ {a, b}, Vector[a × b, Red],
      ViewPoint → {6, 2, 2}]}, Spacings → 100]
```

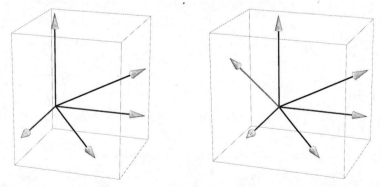

The calculations in the following list verify three of the important properties of the cross product. What are they?

```
{a × b + b × a, a.(a × b), b.(a × b)}
```
$$\{\{0, 0, 0\}, 0, 0\}$$

■ **Example 1.4.1**

Find a unit vector that is perpendicular to each of $\mathbf{a} = \langle 1, 2, 3 \rangle$ *and* $\mathbf{b} = \langle -2, 1, 3 \rangle$.

The solution is straightforward. We simply compute the cross product of the two vectors and then divide by its length.

```
a = {1, 2, 1}; b = {-2, 1, 2};
aCrossb = a × b
n = aCrossb / √aCrossb.aCrossb
N[%]
```

$$\{3, -4, 5\}$$

$$\left\{ \frac{3}{5\sqrt{2}}, -\frac{2\sqrt{2}}{5}, \frac{1}{\sqrt{2}} \right\}$$

$$\{0.424264, -0.565685, 0.707107\}$$

Here's a plot of $\mathbf{a}$, $\mathbf{b}$, and $\mathbf{n}$ as position vectors.

```
Show[Vector[a], Vector[b], Vector[n, Red],
   ijk, Axes → True, ViewPoint → {6, 2, 2}]
```

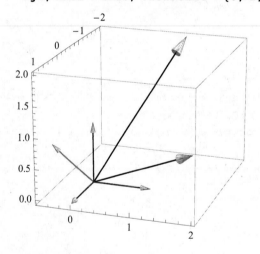

■ **Example 1.4.2**

Find a unit vector that is perpendicular to the plane that passes through the points (2, 3, 2), (1, −1, 1), *and* (3, 1, 0).

The first step is to compute two vectors that are parallel to the plane.

```
a = {1, -1, 1} - {2, 3, 2}
b = {3, -1, 0} - {2, 3, 2}
```

$$\{-1, -4, -1\}$$

$$\{1, -4, -2\}$$

The situation now is exactly as in Example 2.4.1.

```
aCrossb = a × b

n = aCrossb / √aCrossb.aCrossb

N[%]

{4, -3, 8}
```

$$\left\{\frac{4}{\sqrt{89}}, -\frac{3}{\sqrt{89}}, \frac{8}{\sqrt{89}}\right\}$$

```
{0.423999, -0.317999, 0.847998}
```

The following is a plot of **a**, **b**, and **n**, each with initial point (2, 3, 2), so that **a** and **b** lie in the specified plane.

```
p = {2, 3, 2}; Show[Vector[{p, a}], Vector[{p, b}],
   Vector[{p, n}, Red], ijk, ViewPoint → {6, 2, 2}]
```

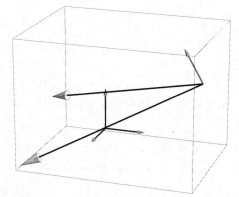

■ Example 1.4.3

Find the area of the triangle whose vertices are the points (3, 1, −1), (7, 5, 1), *and* (3, 3, 3).

The area of the triangle is half that of the parallelogram determined by the vectors

```
a = {7, 5, 1} - {3, 1, -1}
b = {3, 3, 3} - {3, 1, -1}

{4, 4, 2}

{0, 2, 4}
```

The area of the triangle is therefore half the length of the cross product of these vectors.

```
aCrossb = a × b

√aCrossb.aCrossb / 2

N[%]

{12, -16, 8}
```

$$2\sqrt{29}$$

```
10.7703
```

The following plot shows the triangle whose area we've found.

```
p = {3, 1, -1};
Show[Vector /@ {{p, a}, {p, b}, {p + b, a - b}}, ijk, Axes → True]
```

■ Example 1.4.4

Use a cross product to find the area of the triangle in the plane whose vertices are the points $(0, 0)$, (x_1, y_1), *and* (x_2, y_2).

We will simply think of the triangle as having vertices $(0, 0, 0)$, $(x_1, y_1, 0)$, and $(x_2, y_2, 0)$ in three dimensions. The area of the triangle is half that of the parallelogram determined by the vectors

```
Clear[x, y];
a := {x₁, y₁, 0}; b := {x₂, y₂, 0};
```

So the area of the triangle is half the length of the cross product of these vectors.

```
aCrossb = a × b
√ aCrossb.aCrossb / 2
```

$$\{0, 0, -x_2\, y_1 + x_1\, y_2\}$$

$$\frac{1}{2} \sqrt{(-x_2\, y_1 + x_1\, y_2)^2}$$

Note that this result can also be expressed as

$$\frac{1}{2} |x_1\, y_2 - x_2\, y_1| \ .$$

1.5 Equations of Lines

Discussion of the ideas in this section is found in Section 13.5 of Stewart's *Calculus*. (ET:12.5)

Suppose that L is the line in three-dimensional space determined by a point $P_0(x_0, y_0, z_0)$ on the line and a direction vector $\mathbf{v}$. If $\mathbf{r}_0$ is the position vector of P_0 and $\mathbf{r}$ is the position vector of an arbitrary point $P(x, y, z)$ on L, then

$$\mathbf{r} - \mathbf{r}_0 = t\,\mathbf{v}$$

for some scalar t. This is the (vector form of the) parametric equation of the line. The scalar t is the parameter in the equation. Also notice that the equation can be written as

$$\mathbf{r} = \mathbf{r}_0 + t\,\mathbf{v}.$$

For example, consider

```
r = {x, y, z}; r0 = {1 / 2, 2 / 3, 1}; v = {1 / 2, -1 / 2, -1 / 8};
```

so that the equation of the line is

```
r == r0 + t v
```

$$\{x, y, z\} == \left\{ \frac{1}{2} + \frac{t}{2}, \frac{2}{3} - \frac{t}{2}, 1 - \frac{t}{8} \right\}$$

The picture below shows position vectors of several points on the line determined by $\mathbf{r}_0$ and $\mathbf{v}$, as well as a representation of $\mathbf{v}$ with its initial point at $(1/2, 2/3, 1)$.

```
Needs["Vectors`"];
vecs = Append[Table[r0 + t v, {t, -.8, .8, .4}], {r0, v}];
vecplot = Show[Vector /@ vecs, ijk, ViewPoint → {6, 2, 2}]
```

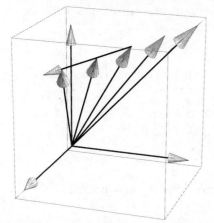

The line in question can be plotted with *Mathematica*'s `ParametricPlot3D` command. (You will become very familiar with `ParametricPlot3D` in the remaining sections of this manual.) The following plots the line and then shows it together with the vectors in the previous plot.

```
theLine = ParametricPlot3D[Evaluate[r0 + t v], {t, -2, 2}];
Show[vecplot, theLine]
```

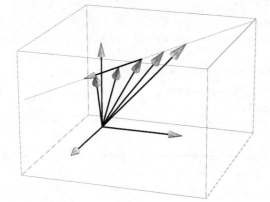

▪ Intersecting Lines

Let's now consider the problem of determining the intersection of two lines,

$$\mathbf{r} = \mathbf{r}_0 + t\,\mathbf{v} \ \text{ and } \ \mathbf{r} = \mathbf{q}_0 + t\,\mathbf{w}.$$

(Note that in three-dimensional space, two such lines need not intersect, even if they are not parallel.)

▪ Example 1.5.1

Determine the intersection of the two lines given by the parametric equations

$$\mathbf{r} = \langle 1 + 2\,t, \ 1 - t/2, \ t/3 \rangle \ \text{ and } \ \mathbf{r} = \langle t, \ 2 - t/2, \ t/9 \rangle.$$

The important thing to realize here is that *the two lines may have a common point that corresponds to different values of the parameter in the two equations.* So the problem becomes that of finding values of t and s such that

$$\langle 1 + 2t, \ 1 - t/2, \ t/3 \rangle = \langle s, \ 2 + s/2, \ s/9 \rangle.$$

Mathematica's `Solve` command will perform the needed calculations as follows.

```
r0 = {1, 1, 0}; q0 = {0, 2, 0};
v = {2, -1 / 2, 1 / 3}; w = {1, -1 / 2, 1 / 9};

Solve[r0 + t v == q0 + s w, {t, s}]

{{t → 1, s → 3}}
```

So we find that the lines intersect at the point given either by $\mathbf{r}_0 + t\,\mathbf{v}$ or by

```
q0 + 3 w
```

$$\left\{ 3, \ \frac{1}{2}, \ \frac{1}{3} \right\}$$

1.6 Equations of Planes

Discussion of the ideas in this section is found in Section 13.5 of Stewart's *Calculus*. (ET:12.5)

A plane in three-dimensional space, determined by a point $P_0(x_0, y_0, z_0)$ on the plane and a normal vector **n**, is described by the vector equation

$$(\mathbf{r} - \mathbf{r}_0) \cdot \mathbf{n} = 0,$$

which, of course, can also be written as $\mathbf{r} \cdot \mathbf{n} = \mathbf{r}_0 \cdot \mathbf{n}$. Notice that even though we refer to this as the *vector* equation of the plane, the quantity on each side of the equation is actually a *scalar*. Expanding the dot products in the equation in terms of the components of each of the vectors involved produces the scalar equation of the plane. With

$$\mathbf{r} = \langle x, y, z \rangle, \mathbf{r}_0 = \langle x_0, y_0, z_0 \rangle, \text{ and } \mathbf{n} = \langle a, b, c \rangle,$$

the equation becomes

$$a(x - x_0) + b(y - y_0) + c(z - z_0) = 0,$$

or

$$ax + by + cz = ax_0 + by_0 + cz_0.$$

One very important thing to notice is that the components of the normal vector become coefficients of the variables in the equation.

Mathematica's `ContourPlot3D` command provides a simple way of plotting a plane.

▪ Example 1.6.1

The following gives us a plot of the plane given by $x - y/2 + 2z = 0$, which is orthogonal to the vector $\mathbf{n} = \langle 1, -1/2, 2 \rangle$ and passes through the origin. It also displays the normal vector **n**. (Notice that the vectors **i** and **k** are in front of the plane, while **j** is behind it.)

```
n = {1, -1 / 2, 2}; r = {x, y, z}; r0 = {0, 0, 0}; Needs["Vectors`"];
Show[ContourPlot3D[(r - r0).n == 0, {x, -2, 2}, {y, -2, 2},
   {z, -2, 2}, ContourStyle → Opacity[.9], Mesh → None],
Vector[n], ijk, ViewPoint → {6, 2, 2}]
```

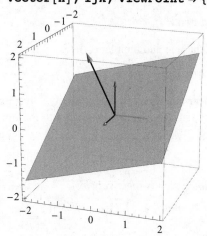

A non-vertical plane can be plotted with the `Plot3D` command as well. This requires solving the equation for z and then plotting the result as a function of two variables.

■ Example 1.6.2

The plane through the point $(0, 1/2, -1)$ with normal vector $\mathbf{n} = \langle 1, 1/2, 2 \rangle$ can be plotted with `Plot3D` as follows.

```
n = {1 / 3, 1 / 3, 1}; r = {x, y, z}; r0 = {1 / 4, 1 / 4, 0};
f[x_, y_] = z /. Flatten[Solve[(r - r0).n == 0, z]]
```

$$\frac{1}{6} (1 - 2 x - 2 y)$$

```
Show[Plot3D[f[x, y], {x, -2, 2},
    {y, -1.5, 1.5}, Mesh → None, PlotStyle → Opacity[.9]],
  Vector[{r0, n}], ijk, ViewPoint → {6, 2, 2},
    Axes → False, BoxRatios → Automatic, PlotRange → All]
```

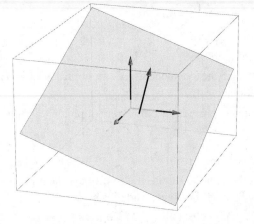

■ Intersection of a Line and a Plane

Suppose we wish to determine the intersection of a line and a plane given respectively by

$$\mathbf{r} = \mathbf{r}_0 + t\,\mathbf{v} \ \text{ and } \mathbf{r} \cdot \mathbf{n} = \mathbf{q}_0 \cdot n.$$

If $\mathbf{r}$ is the position vector of a point common to the line and the plane, then

$$(\mathbf{r}_0 + t\,\mathbf{v}) \cdot \mathbf{n} = \mathbf{q}_0 \cdot \mathbf{n}$$

for some value of the parameter t. This is a linear equation in t that is easily solved, so long as $\mathbf{v}$ and $\mathbf{n}$ are not orthogonal. Once t is known, substitution of its value into the line equation reveals the intersection point.

■ Example 1.6.3

Find the point of intersection of the line and plane given respectively by

$$\langle x, y, z \rangle = \langle t, 1 - t, \ t - 1/3 \rangle \ \text{ and } \ 2x - y + 3z = 3.$$

For convenience, let's first define vectors

```
r = {x, y, z}; r0 = {0, 1, -1/3};
v = {1, -1, 1}; n = {2, -1, 3};
```

and functions

```
lineF[t_] = r0 + t v
planeF[{x_, y_, z_}] = r.n
```

$$\left\{t, 1 - t, -\frac{1}{3} + t\right\}$$

$$2x - y + 3z$$

We'll substitute the *t*-expressions for *x*, *y*, and *z* from the line equation into the left-hand side of the plane equation by composing the two functions above.

```
planeF[lineF[t]] // Simplify
```

$$-2 + 6t$$

Now we solve the plane equation for *t*,

```
Solve[% == 3, t] // Flatten
```

$$\left\{t \to \frac{5}{6}\right\}$$

and finally substitute the found *t* into the line equation:

```
p0 = lineF[t] /. %
```

$$\left\{\frac{5}{6}, \frac{1}{6}, \frac{1}{2}\right\}$$

So the point of intersection is $(5/6, 1/6, 1/2)$.

It is worth looking here at a slightly different, more straightforward, method. The problem at hand really just comes down to solving a system of four equations in four unknowns *t*, *x*, *y*, and *z*.

```
planeEQ = planeF[{x, y, z}] == 3; lineEQ = r == lineF[t];
system = Append[{planeEQ}, lineEQ]
```

$$\left\{2x - y + 3z == 3, \{x, y, z\} == \left\{t, 1 - t, -\frac{1}{3} + t\right\}\right\}$$

We'll use `Solve` to find the solution of the system and then pick out the values of *x*, *y*, and *z*.

```
Solve[system, {t, x, y, z}] // Flatten
p0 = {x, y, z} /. %
```

$$\left\{t \to \frac{5}{6}, x \to \frac{5}{6}, y \to \frac{1}{6}, z \to \frac{1}{2}\right\}$$

$$\left\{\frac{5}{6}, \frac{1}{6}, \frac{1}{2}\right\}$$

The following is a plot of the line and plane in question, together with the line's direction vector **v** (with the solution as its initial point) and the plane's normal vector (with initial point $(0, 0, 1)$).

```
Show[ParametricPlot3D[Evaluate[r0 + t v], {t, -1, 3}],
  ContourPlot3D[r.n - 3 == 0, {x, -2, 2}, {y, -2, 2},
    {z, -2, 4}, Mesh → None, ContourStyle → Opacity[.75]],
  Vector /@ {{{0, 0, 1}, n}, {p0, v}}, ijk,
    PlotRange → All, ViewPoint → {6, 2, 2}]
```

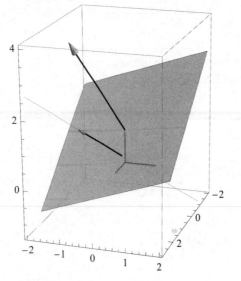

■ Intersection of Two Planes

The problem of interest here is that of finding the intersection of two planes:

$$\mathbf{r} \cdot \mathbf{n}_1 = \mathbf{r}_0 \cdot \mathbf{n}_1 \quad \text{and} \quad \mathbf{r} \cdot \mathbf{n}_2 = \mathbf{q}_0 \cdot \mathbf{n}_2$$

If the two normal vectors are not parallel, the planes must intersect in a line. If the two normal vectors are parallel, the planes will be parallel, and so the intersection will be empty unless the two planes are identical.

■ Example 1.6.4

Determine the intersection of the two planes given by

$$2x + z = 1 \quad \text{and} \quad -x - y - 2z = 0.$$

There are various ways to approach this problem. One way is to observe that:

• The normal vectors are not parallel; therefore the planes are not parallel; therefore the intersection of the two planes must be line.

• Any direction vector of the line of intersection must be orthogonal to the normal vector of each of the two planes.

Consequently, the vector equation of the line of intersection can be determined by finding one point of intersection P_0 and a direction vector **v**.

To find a point of intersection, we can solve the two plane equations for two variables, say x and y, in terms of the third and then substitute some number for that third variable. Here is the result of using `Solve` to find x and y in terms of z.

```
r = {x, y, z}; n1 = {2, 0, 1}; n2 = {-1, -1, -2};
soln = Solve[{r.n1 == 1, r.n2 == 0}, {x, y}] // Flatten
```

$$\left\{x \to \frac{1-z}{2}, \; y \to \frac{1}{2}\,(-1-3\,z)\right\}$$

Now we'll substitute $z = 0$ and arrive at a point common to both planes.

```
r0 = {x, y, z} /. soln /. z → 0 // Flatten
```

$$\left\{\frac{1}{2}, \; -\frac{1}{2}, \; 0\right\}$$

The next step is to find a direction vector for the line of intersection. The key here is that such a vector must be orthogonal to each of the normal vectors of the two planes. Such a vector is the cross product of the two normal vectors.

```
v = n1 × n2
```

$$\{1, 3, -2\}$$

Now we have everything we need to construct the equation of the line in which the two planes intersect.

```
r == r0 + t v
```

$$\{x, y, z\} == \left\{\frac{1}{2} + t, \; -\frac{1}{2} + 3\,t, \; -2\,t\right\}$$

The following plots the two planes, the line of intersection, and the line's direction vector.

```
Show[ContourPlot3D[(r - r0).n1 == 0, {x, -2, 2}, {y, -2.5, 1.8},
   {z, -2, 2}, Mesh → None, ContourStyle → Opacity[.75]],
  ContourPlot3D[(r - r0).n2 == 0, {x, -1.5, 2.2}, {y, -2.5, 1.8},
   {z, -2, 2}, Mesh → None, ContourStyle → Opacity[.75]],
  Vector[{r0, v}], ParametricPlot3D[Evaluate[r0 + t v], {t, -1, 1.2}],
  PlotRange → All, BoxRatios → Automatic, ViewPoint → {4, 3, 2}]
```

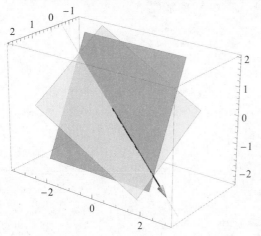

◆ Exercises

13. Find a unit vector that is perpendicular to each of the vectors

$$\mathbf{a} = \langle 1, 2, 3 \rangle \text{ and } \mathbf{b} = \langle 3, 2, 1 \rangle.$$

14. Find a unit vector that is perpendicular to the plane that passes through the points $(1, 1, 1)$, $(1, 2, 2)$, and $(3, -1, 0)$.

15. Find the area of the triangle whose vertices are the points $(1, 2, 3)$, $(3, 2, 1)$, and $(2, 1, 3)$.

16. Determine the intersection of the following pairs of lines. Describe each result geometrically.

    a) $\mathbf{r} = \langle 1 - t, \ 1 + 2t, \ t \rangle$ and $\mathbf{r} = \langle 2 - 3t, \ 5t - 1/2, \ 1 - t \rangle$

    b) $\mathbf{r} = \langle 1 - t, \ 1 + 2t, \ t \rangle$ and $\mathbf{r} = \langle 2 - 3t, \ 5t - 1/2, \ 1 + t \rangle$

    c) $\mathbf{r} = \langle 1 - t, \ 1 + 2t, \ t \rangle$ and $\mathbf{r} = \langle t - 1, \ 5 - 2t, \ 2 - t \rangle$

17. Find a parametric equation for the line that passes through the point $(1, 2, -1)$ and is perpendicular to each of the vectors $\mathbf{a} = \langle 1, 1, 2 \rangle$ and $\mathbf{b} = \langle 2, -2, 1 \rangle$.

18. Find a parametric equation for the line that passes through the point $(3, -2, 1)$ and is perpendicular to each of the lines

$$\mathbf{r} = \langle 1 - t, \ 1 + 2t, \ t \rangle \text{ and } \mathbf{r} = \langle 2 - 3t, \ 5t - 1/2, \ 1 - t \rangle.$$

19. Find the point of intersection of the line and plane given respectively by

$$\langle x, \ y, \ z \rangle = \langle 2t, 1 - t, \ t \rangle \text{ and } 3x - 2y + z = 5.$$

20. Find the point where the line passing through $(5, 3, 1)$ and $(1, 1, 2)$ intersects the plane given by $3x + y + z = 3$.

21. Determine the intersection of the two planes given by

$$x - y + z = 2 \text{ and } 2x + y - 2z = 0.$$

22. Find the equation of the plane that passes through the point $(3, 2, 1)$ and is perpendicular to the line given by $\mathbf{r} = \langle 1 - t, \ 1 + 2t, \ t \rangle$.

2 Surfaces

One of the most important, useful, and interesting abilities that *Mathematica* gives us is visualization of surfaces in three-dimensional space. Such a surface may be the *graph* of a function of two variables or a *level surface* (or *contour*) of a function of three variables.

2.1 Graphs of Functions of Two Variables

> Functions of two variables are discussed in Section 15.1 of Stewart's *Calculus*. (ET:14.1)

The `Plot3D` command graphs functions of two variables. Various options may be given to affect the appearance of the plot. We will illustrate a few of these options in the following examples. You can also rotate three-dimensional plots by clicking and dragging your mouse.

■ **Example 2.1.1**

The graph of the function $f(x, y) = -(x^2 + y^2)/2$ (or of the equation $x^2 + y^2 + 2z = 0$) for $-2 \le x \le 2$ and $-2 \le y \le 2$ is plotted with no options as follows.

```
f[x_, y_] := - (x^2 + y^2) / 2; Plot3D[f[x, y], {x, -2, 2}, {y, -2, 2}]
```

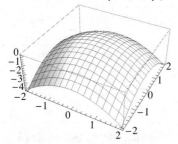

♡ **The BoxRatios and Axes options**. If we want identical scales on each of the three axes, we can specify `BoxRatios→Automatic`. This gives a much more dramatic picture and is seen below. To reduce clutter in the plot we also set `Axes→False`.

```
Plot3D[f[x, y], {x, -2, 2}, {y, -2, 2},
    BoxRatios → Automatic, Axes → False]
```

We can also specify the shape of the "bounding box" with the `BoxRatios` option.

```
Plot3D[f[x, y], {x, -2, 2}, {y, -2, 2},
  BoxRatios → {3, 1, 2}, Axes → False]
```

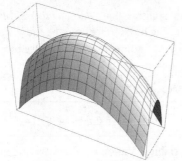

To dispense with the box and axes altogether, specify `Axes→False` and `Boxed→False`.

```
Plot3D[f[x, y], {x, -2, 2}, {y, -2, 2},
  BoxRatios → Automatic, Axes → False, Boxed → False]
```

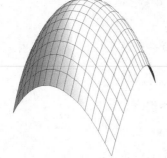

■ The 3D View Point

Mathematica's default axis orientation is seen in the following plot of the vectors **i**, **j**, and **k**. The y-axis points "into" the page and slightly to the right. Imagine looking down the z-axis from the front side of the page, facing slightly toward the right. From that viewpoint, the xy-plane looks the way we are accustomed to seeing it.

```
Needs["Vectors`"]; Show[ijk, Axes → True, AxesLabel → {x, y, z}]
```

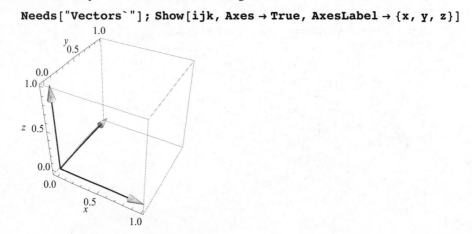

☿ The `ViewPoint` option allows us to change the axis orientation and thereby view surfaces from different angles. For instance, with `ViewPoint→{6,2,2}` the axes are oriented in more-or-less the way we're use to drawing them by hand, with the *x*-axis pointing "out" of the page and a bit to the left.

Show[ijk, ViewPoint → {6, 2, 2}, Axes → True, AxesLabel → {x, y, z}]

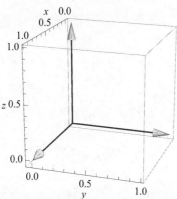

■ Example 2.1.2

This plots $f(x, y) = x^2 - y^2$ with `ViewPoint→{6,2,2}`:

Plot3D[f[x, y], {x, -1, 1}, {y, -1, 1}, BoxRatios → Automatic,
    ViewPoint → {6, 2, 2}, AxesLabel → {x, y, z}]

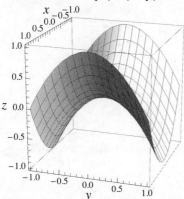

The following displays a `GraphicsRow` containing several different views of the surface.

GraphicsRow[Table[Plot3D[f[x, y], {x, -1.5, 1.5}, {y, -1.5, 1.5},
    BoxRatios → Automatic, ViewPoint → {3 Cos[k π], 3 Sin[k π], 1},
    Axes → False], {k, 0, .8, .2}], Spacings → 0]

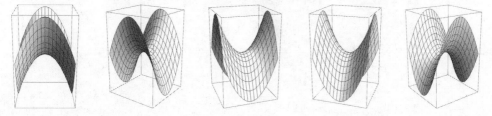

■ Example 2.1.3

The two surfaces we used as examples so far were easy surfaces for *Mathematica* to plot nicely, even with only default options. A more difficult example is

```
f[x_, y_] := Cos[x y] - Sin[x y];
Plot3D[f[x, y], {x, -5, 5}, {y, -5, 5}]
```

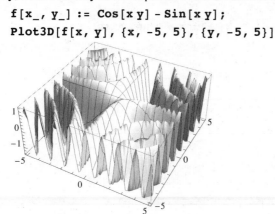

Notice that the result is somewhat "jaggy." To get a smoother plot, we'll set MaxRecursion→4.

```
Plot3D[f[x, y], {x, -4, 4}, {y, -4, 4}, MaxRecursion → 4, Axes → None]
```

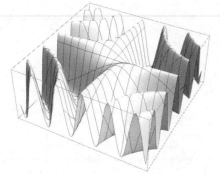

It is sometimes desirable to remove the grid curves, or mesh, from the surface by setting Mesh→None.

```
Plot3D[f[x, y], {x, -4, 4}, {y, -4, 4},
  MaxRecursion → 4, Axes → None, Mesh → None]
```

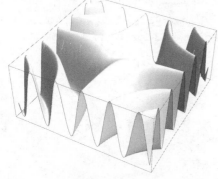

Be aware that increasing `MaxRecursion` slows the plotting of the graph and uses more memory. Generally speaking, you'll find that in most cases the useful range for `MaxRecursion` is roughly 2 to 6.

■ Example 2.1.4

Another issue that sometimes arises involves *Mathematica*'s occasional tendency to clip, or chop off, the top or bottom of a surface. Here's an example:

```
f[x_, y_] := Exp[-x^2 - y^2];
Plot3D[f[x, y], {x, -4, 4}, {y, -4, 4}, Axes → {False, False, True}]
```

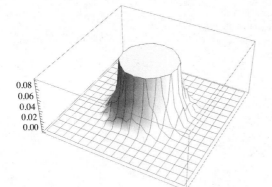

Notice the flat portion in the center of the surface. This is not a feature of the correct graph.

♡ **The `PlotRange` option**. An easy fix to the clipping problem is to set `PlotRange→All`.

```
Plot3D[f[x, y], {x, -4, 4}, {y, -4, 4},
 PlotRange → All, Axes → {False, False, True}]
```

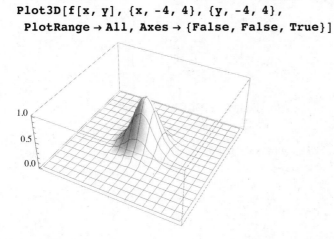

2.2 Traces and Contours

Traces are discussed in Section 13.6 of Stewart's *Calculus*. Contours (level curves) are discussed in Section 15.1. (ET: 12.6 & 14.1)

To understand the graph of a equation in three variables x, y, and z, it is often helpful to look at cross-sections obtained by setting one of the three variables equal to a constant and graphing the resulting equation in the other two variables. Such cross-sections are examples of *traces* of the graph.

■ Example 2.2.1

To illustrate traces, let's consider the equation $x^2 + y^3 - 2z = 0$. Note that the graph of this equation is the same as the graph of the function $f(x, y) = \frac{1}{2}\left(x^2 + y^3\right)$.

```
f[x_, y_] := .5 (x^2 + y^3); Plot3D[f[x, y], {x, -1, 1},
  {y, -1, 1}, BoxRatios → Automatic, ViewPoint → {6, 2, 2}]
```

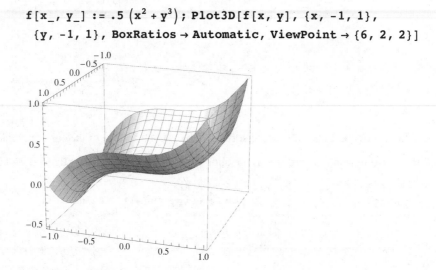

Setting x to a constant amounts to slicing the surface vertically, parallel to the yz-plane. Traces in the vertical planes $x = k$ are the cubic curves $z = \frac{1}{2}\left(k^2 + y^3\right)$.

```
xtraces = Table[f[k, y], {k, 0, 5}];
Plot[Evaluate[xtraces], {y, -3, 3}]
```

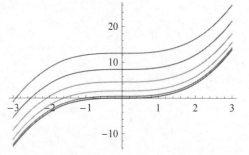

Setting y to a constant amounts to slicing the surface vertically, parallel to the xz-plane. Traces in the vertical planes $y = k$ are the parabolas $z = \frac{1}{2}\left(x^2 + k^3\right)$.

```
xtraces = Table[f[x, k], {k, -2, 2}];
Plot[Evaluate[xtraces], {x, -4, 4}]
```

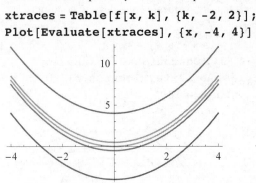

Setting z to a constant amounts to slicing the surface horizontally, parallel to the xy-plane. The traces in horizontal planes $z = k$ are the curves $2k = x^2 + y^3$, or $y = \sqrt[3]{2k - x^2}$.

```
xtraces = Table[∛(Abs[2 k - x²]) Sign[2 k - x²], {k, -2, 3}];
Plot[Evaluate[xtraces], {x, -3, 3}]
```

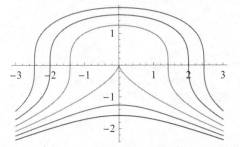

$\heartsuit$ **ContourPlot.** Horizontal traces of graph of a function $f(x, y)$ are especially important and are usually called **contours**, or **level curves**, of the function f. *Mathematica* has a function called ContourPlot for plotting these curves. In the present example (with no options) it works like this:

```
ContourPlot[f[x, y], {x, -1, 1}, {y, -1, 1}]
```

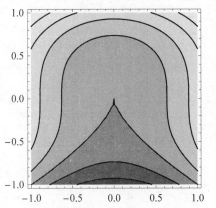

Useful options for `ContourPlot` include `Contours`, `ContourShading`, `Contour-Style`, `Frame`, and `PlotPoints`. The `Contours` option can be used to specify either the number of contours to be plotted or a list of z-values for the desired contours.

```
pic1 = ContourPlot[f[x, y], {x, -1, 1}, {y, -1, 1}, Contours → 15];
pic2 = ContourPlot[f[x, y], {x, -2, 2}, {y, -2, 2},
    Contours → {-1, 0, .5, 1, 2, 3}, ContourShading → False,
    Frame → False, ContourStyle → AbsoluteThickness[1]];
GraphicsRow[{pic1, pic2}, Spacings → 100]
```

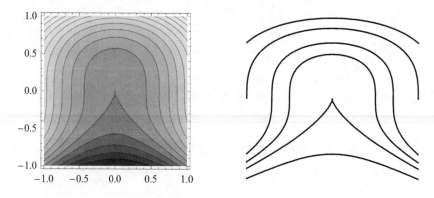

◆ Exercises

For each function in Exercises 1-10 below, obtain a good plot of its graph with `Plot3D` by choosing appropriate variable ranges and specifying appropriate options. Then create a `GraphicsRow` containing a contour plot and plots of traces in $x = k$ and $y = k$. An example of the desired result is shown for Exercise 1.

1. $f(x, y) = \sin(2x + y) + \cos(x - y)$

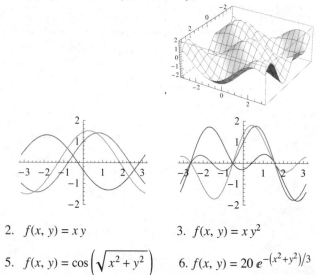

2. $f(x, y) = xy$

3. $f(x, y) = xy^2$

4. $f(x, y) = \cos(x + \sin y)$

5. $f(x, y) = \cos\left(\sqrt{x^2 + y^2}\right)$

6. $f(x, y) = 20\, e^{-(x^2+y^2)/3}$

7. $f(x, y) = x\cos y + y\sin x$

8. $f(x, y) = \cos x \, y - \cos x \cos y$ 9. $f(x, y) = e^{-(x^2+y^2)/4} \cos\left(4\sqrt{x^2+y^2}\right)$

10. $f(x, y) = \left(5xy - 2x^3 - 3y^3\right)\big/\left(1 + x^4 + y^4\right)$

2.3 Cylindrical and Spherical Coordinates

> Cylindrical and spherical coordinates are discussed in Section 13.7 of Stewart's *Calculus*. (ET:12.7)

If is often more convenient to describe certain surfaces in terms of coordinate systems other than the usual rectangular "*xyz*" coordinates. Two of the most important alternative coordinate systems are **cylindrical coordinates** and **spherical coordinates**.

■ Cylindrical Coordinates

Cylindrical coordinates (r, θ, z) are related to rectangular coordinates by

$$x = r \cos \theta, \quad y = r \sin \theta, \quad \text{and} \quad z = z.$$

These formulas describe the conversion from cylindrical coordinates to rectangular coordinates. It will be useful to incorporate them into a function:

```
cylToRect[{r_, θ_, z_}] := {r Cos[θ], r Sin[θ], z}
```

For example, the point whose cylindrical coordinates are $(1, \pi/3, 2)$ has rectangular coordinates given by

```
cylToRect[{1, π / 3, 2}]
```

$$\left\{\frac{1}{2}, \frac{\sqrt{3}}{2}, 2\right\}$$

Let's now look at a few graphs of equations in cylindrical coordinates. `Parametric-Plot3D` plots the graph of a list of three parametric equations involving one or two parameters. For equations in cylindrical coordinates, those parametric equations are

$$x = r \cos \theta, \quad y = r \sin \theta, \quad \text{and} \quad z = z,$$

that is,

```
{x, y, z} == cylToRect[{r, θ, z}]
```

$\{x, y, z\} == \{r \, Cos[\theta], r \, Sin[\theta], z\}$

where one of the variables r, θ, and z is expressed in terms of the other two. Combining the `ParametricPlot3D` command with our `cylToRect` function makes it easy to plot surfaces described with cylindrical coordinates, as seen in the following examples.

- ## Example 2.3.1

A very simple equation in cylindrical coordinates is $r = 1$, which describes the cylinder of radius 1, centered on the z-axis. The portion of this cylinder between $z = 0$ and $z = 2$ is plotted as follows.

```
fn = cylToRect[{1, θ, z}];
ParametricPlot3D[fn, {θ, 0, 2 π}, {z, 0, 2}]
```

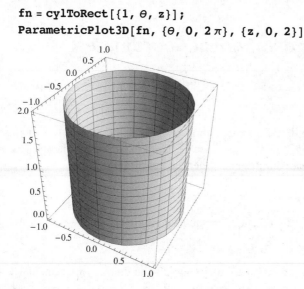

- ## Example 2.3.2

The equation $z = r$, which is equivalent to $z^2 = x^2 + y^2$ in rectangular coordinates, is plotted as follows for $-2 \leq z \leq 2$.

```
ParametricPlot3D[cylToRect[{r, θ, r}], {r, -2, 2}, {θ, 0, 2 π}]
```

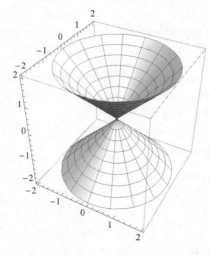

- ## Example 2.3.3

The equation $z = r^2$, which is equivalent to $z = x^2 + y^2$ in rectangular coordinates, is plotted as follows for $0 \le r \le 2$.

```
ParametricPlot3D[cylToRect[{r, θ, r²}], {r, 0, 2}, {θ, 0, 2π}]
```

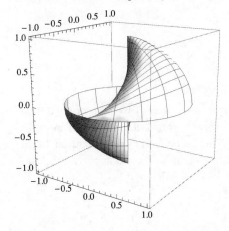

- ## Example 2.3.4

A rather peculiar surface is given by $z = \sqrt{1 - r^2}\, \cos\theta$, which, for $0 \le r \le 1$, is the cylindrical version of

$$z = x \sqrt{\frac{1 - x^2 - y^2}{x^2 + y^2}}\,.$$

The portion of the surface with $0 \le r \le 1$ is plotted as follows.

```
fn = cylToRect[{r, θ, √(1 - r²) Cos[θ]}];
ParametricPlot3D[fn, {r, 0, 1}, {θ, 0, 2π}, ViewPoint → {2, -3, 1}]
```

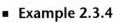

■ Spherical Coordinates

Spherical coordinates (ρ, θ, ϕ) are related to rectangular coordinates by

$$x = \rho \sin \phi \cos \theta, \ \ y = \rho \sin \phi \sin \theta, \ \text{and} \ \ z = \rho \cos \phi.$$

The following defines a function that converts spherical coordinates to rectangular coordinates.

```
sphrToRect[{ρ_, θ_, φ_}] := {ρ Sin[φ] Cos[θ], ρ Sin[φ] Sin[θ], ρ Cos[φ]}
```

For example, the point whose spherical coordinates are $(1, \pi/3, \pi/6)$ has rectangular coordinates given by

```
sphrToRect[{1, π / 3, π / 6}]
```

$$\left\{ \frac{1}{4}, \ \frac{\sqrt{3}}{4}, \ \frac{\sqrt{3}}{2} \right\}$$

Let's now look at a few graphs of equations in spherical coordinates. Just as we combined `ParametricPlot3D` with our `cylToRect` function to plot cylindrical-coordinate surfaces, we will combine `ParametricPlot3D` with our `sphrToRect` function to plot surfaces described with spherical coordinates.

■ Example 2.3.5

A very simple example of an equation in spherical coordinates is $\rho = 1$, which describes the sphere of radius 1, centered at the origin.

```
sphr = sphrToRect[{1, θ, φ}]; ParametricPlot3D[sphr,
   {θ, 0, 2 π}, {φ, 0, π}, Axes → False, Boxed → False]
```

It is very instructive to look at what happens to the sphere if we restrict the domains of the angles θ and ϕ. The first of the following two plots shows the portion of the sphere corresponding to $0 \le \theta \le 7\pi/4$ and $0 \le \phi \le \pi$. The second shows the portion of the sphere corresponding to $0 \le \theta \le 2\pi$ and $\pi/4 \le \phi \le \pi$. Notice how the domains of the angles relate to missing parts of the sphere.

```
GraphicsRow[{ ParametricPlot3D[sphr,
    {θ, 0, 7 π / 4}, {φ, 0, π}, Boxed → False, Axes → False],
  ParametricPlot3D[Evaluate[sphr], {θ, 0, 2 π},
    {φ, π / 4, π}, Boxed → False, Axes → False]}]
```

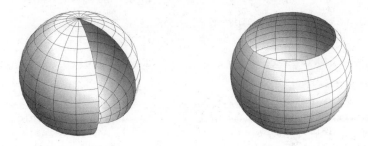

- ## Example 2.3.6

A "clam shell" surface is described in spherical coordinates by

$$\rho = \theta, \ 0 \le \theta \le 3\pi.$$

Note that for fixed values of ϕ, grid curves will spiral one and one-half times around the z-axis. For each fixed value of θ, the corresponding grid curve will be a circle that lies in a vertical plane containing the z-axis. The following shows the portion of the surface corresponding to $0 \le \theta \le 3\pi$ for each of the intervals $0 \le \phi \le \pi$ and $0 \le \theta \le 11\pi/2$.

```
GraphicsRow[{ParametricPlot3D[sphrToRect[{θ, θ, φ}],
    {θ, 0, 3 π}, {φ, 0, π}, ViewPoint → {1, -3, 1}],
  ParametricPlot3D[sphrToRect[{θ, θ, φ}],
    {θ, 0, 3 π}, {φ, π / 2, π}, Mesh → {15, 7},
    PlotRange → {All, All, {-3 π, 3 π}}, ViewPoint → {1, -3, 1}]}]
```

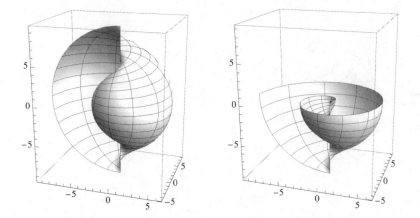

■ Example 2.3.7

To see a few more nice spherical-coordinate plots, let's create a `GraphicsGrid` containing the graph of each of the equations

$$z = \phi, \quad \rho = \cos(3\phi/4), \quad \rho = 1 + \sin\phi, \text{ and } \phi = 1 + \sin\rho.$$

```
{fn1, fn2, fn3, fn4} = sphrToRect/@
    {{ϕ, θ, ϕ}, {Cos[3 ϕ/4], θ, ϕ}, {1 + Sin[ϕ], θ, ϕ}, {ρ, θ, 1 + Sin[ρ]}};

opts := {Axes→False, Boxed→False, PlotRange→All, ViewPoint→{3,1,1}}

GraphicsGrid[{
  {ParametricPlot3D[fn1, {θ, 0, 2 π}, {ϕ, -π/3, π/3}, Evaluate[opts]],
   ParametricPlot3D[fn1, {θ, π/4, 2 π}, {ϕ, 0, π}, Evaluate[opts]]},
  {ParametricPlot3D[fn3, {θ, π/4, 2 π}, {ϕ, 0, 2 π}, Evaluate[opts]],
   ParametricPlot3D[fn4, {θ, 0, 2 π}, {ρ, -π/2, π/5}, Evaluate[opts]]}
  }, Spacings → {0, -50}]
```

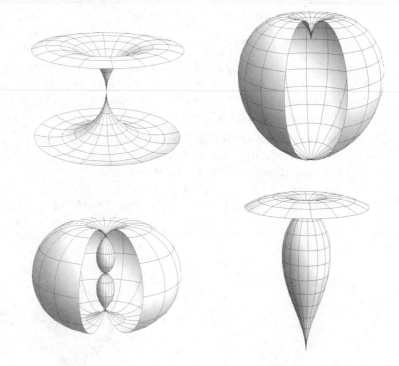

2.4 3D Contours and the Quadric Surfaces

Quadric surfaces are discussed in Section 13.6 of Stewart's *Calculus*. (ET:12.6)

We have seen that the contours of a function f of two variables are the family of plane curves given by the equation $f(x, y) = k$ as k takes on fixed values in the range of f. We can interpret these contours as horizontal cross-sections of the graph of f.

Just as the contours of a function of two variables are curves in two dimensions, the contours of a function of three variables are surfaces in three dimensions corresponding to the graphs of the equations $f(x, y, z) = k$. We can think of these contours as cross-sections of the four-dimensional graph of the function, which of course is quite difficult to visualize!

Mathematica's `ContourPlot3D` command plots three-dimensional contours.

■ Example 2.4.1

A simple example is the unit sphere, whose equation in rectangular coordinates is $x^2 + y^2 + z^2 = 1$. This sphere is a contour corresponding to $k = 1$ of the function

```
f[x_, y_, z_] := x^2 + y^2 + z^2
ContourPlot3D[f[x, y, z] == 1, {x, -1, 1}, {y, -1, 1}, {z, -1, 1}]
```

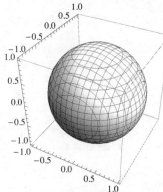

A slight variation results in an **ellipsoid**:

```
ContourPlot3D[ (x / 3)^2 + y^2 + z^2 == 1, {x, -3, 3},
  {y, -1, 1}, {z, -1, 1}, BoxRatios → Automatic,
  AxesEdge → {Automatic, Automatic, {-1, -1}}]
```

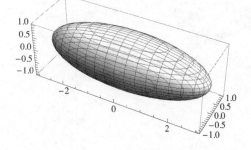

■ Quadric Surfaces

Spheres and ellipses are examples of **quadric surfaces**. Quadric surfaces are contours of quadratic functions in three variables—or, equivalently, graphs of quadratic equations in three variables. When one of the three variables appears to only the first power, it is straightforward to solve for that variable and plot the surface with `Plot3D`. These plots are typical. The first is a *hyperbolic paraboloid* (or *saddle*); the second is an *elliptical paraboloid*.

```
GraphicsRow[{
    Plot3D[x^2 - y^2, {x, -1, 1}, {y, -1, 1}, BoxRatios → Automatic],
    Plot3D[x^2 / 4 + y^2, {x, -4, 4}, {y, -2, 2}, BoxRatios → Automatic]}]
```

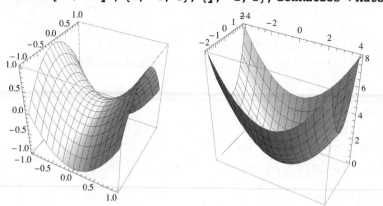

It is also possible to piece together any quadric surface from two applications of `Plot3D` in general, but `ContourPlot3D` provides a convenient way to plot a quadric surface with one command.

■ Example 2.4.2

The following are typical quadric surfaces corresponding to equations that are quadratic in all three variables. The first is a *hyperboloid of one sheet*; the second is a *cone*; the third is a *hyperboloid of two sheets*.

```
GraphicsRow[
   {ContourPlot3D[x^2 + y^2 - z^2 == 1, {x, -4, 4}, {y, -4, 4}, {z, -3, 3}],
    ContourPlot3D[x^2 + y^2 - z^2 == 0, {x, -1, 1}, {y, -1, 1}, {z, -1, 1}],
    ContourPlot3D[x^2 + y^2 - z^2 == -1, {x, -4, 4}, {y, -4, 4}, {z, -4, 4}]}]
```

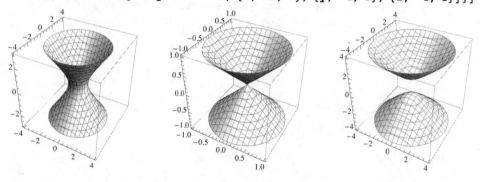

◆ Exercises

11. Plot the paraboloid $z = x^2 + y^2$ for $-2 \le x \le 2$ and $-2 \le y \le 2$ using `Plot3D` (with `BoxRatios→Automatic`). Then use `ParametricPlot3D` to plot the graph of the equivalent cylindrical-coordinate equation $z = r^2$ for $0 \le r \le 2$ and $0 \le \theta \le 2\pi$. Discuss the difference in the two plots.

12. Plot the top half of the unit sphere in each of the following four ways. Then discuss the differences between the plots, particularly with regard to the grid curves in each plot.

    a) Use `Plot3D` to plot $z = \sqrt{1 - x^2 - y^2}$ for $-1 \le x \le 1$ and $-1 \le y \le 1$.

    b) Use `ParametricPlot3D` to plot the cylindrical-coordinate equation $z = \sqrt{1 - r^2}$ for $0 \le r \le 1$ and $0 \le \theta \le 2\pi$.

    c) Use `ParametricPlot3D` to plot the spherical-coordinate equation $\rho = 1$ for $0 \le \theta \le 2\pi$ and $0 \le \phi \le \pi/2$.

    d) Use `ContourPlot3D` to plot the top half of an appropriate contour of the function $f(x, y, z) = x^2 + y^2 + z^2$.

13. Use `ParametricPlot3D` to plot the portion of the unit sphere given in spherical coordinates by $\rho = 1$ for each of the following angle ranges. Then experiment with others.

    a) $0 \le \theta \le 2\pi$ and $0 \le \phi \le 2\pi/3$

    b) $0 \le \theta \le 2\pi$ and $\pi/3 \le \phi \le \pi$

    c) $0 \le \theta \le \pi/2$ and $0 \le \phi \le \pi$

    d) $0 \le \theta \le 3\pi/2$ and $0 \le \phi \le \pi$

    e) $0 \le \theta \le \pi/2$ and $0 \le \phi \le \pi/2$

14. The graph of the cylindrical coordinate equation $r = k$ (where k is a constant) is a cylinder centered along the z-axis (if $k \mathbin{!=} 0$). What type of graphs do the equations $z = k$ and $\theta = k$ have? Use `ParametricPlot3D` to plot graphs of $z = 1, 2, 3$ and $\theta = 1, 2, 3$.

15. The graph of the spherical coordinate equation $\rho = k$ (where k is a constant) is a sphere centered at the origin (if $k \ne 0$). What type of graphs do the equations $\theta = k$ and $\phi = k$ have? Use `ParametricPlot3D` to plot graphs of $\theta = 1, 2, 3$ and $\phi = 1, 2, 3$.

3 Vector-valued Functions

Vector functions and space curves are the subject of Chapter 14 in Stewart's *Calculus*. (ET:13)

In this chapter we make extensive use of `Vector`, which is defined in the `Vectors` package. So be sure that you load that package before entering any of the following commands.

```
<< Vectors`
```

3.1 Space Curves

In two separate contexts, you have already encountered vector-valued functions. The first was the parametric equations of a curve in the plane,

$$x = f(t), \quad y = g(t),$$

which can be viewed as the components of a vector-valued function

$$\gamma(t) = \langle f(t), g(t) \rangle.$$

- ## Example 3.1.1

The following is a plot of the curve given by

$$x = \sin(t + \pi/4), \quad y = \cos t$$

together with a sampling of position vectors of points on the curve.

```
γ[t_] := {Sin[t + π / 4], Cos[t]};
Show[ParametricPlot[γ[t], {t, 0, 2 π}],
  Table[Vector[γ[t]], {t, π / 6, 2 π, π / 6}]]
```

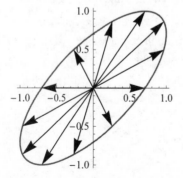

Enter the following, and use the slider to trace out the curve with position vectors.

```
Manipulate[Show[Vector[γ[t]],
  ParametricPlot[γ[s], {s, 0, t}, PlotStyle → {Thick, Red}],
  PlotRange → 1.1 {{-1, 1}, {-1, 1}}, Axes → True], {t, .01, 2 π}]
```

The other context in which you've already seen vector-valued functions is the parametric description of lines in three dimensions. Recall that

$$\mathbf{r} = \mathbf{r}_0 + t\,\mathbf{v}$$

is an equation of the line that is parallel to $\mathbf{v}$ and passes through the point whose position vector is $\mathbf{r}_0$. (See Section 1.5.)

Our main concern here will be functions of one variable whose values are three-dimensional vectors:

$$\gamma(t) = \langle f(t),\, g(t),\, h(t)\rangle = f(t)\,\mathbf{i} + g(t)\,\mathbf{j} + h(t)\,\mathbf{k}.$$

The set of points $(f(t),\, g(t),\, h(t))$, where t varies over the domain of $\gamma(t)$, is called a **space curve**. In other words, the space curve defined by $\gamma(t)$ is the set of points described by the *parametric equations*

$$x = f(t), \quad y = g(t), \quad z = h(t).$$

Likewise, we say that the function $\gamma(t)$ is a *parametrization* of the space curve.

Mathematica's `ParametricPlot3D` command lets us plot space curves.

■ Example 3.1.2

An interesting space curve is traced by

```
γ[t_] := {Sin[t], Sin[2 t], Sin[3 t]}
```

Note that the least common period of the three component functions is 2π; consequently, the entire curve will be traced out exactly once as t varies from 0 to 2π. The following shows the curve along with $\mathbf{i}$, $\mathbf{j}$, $\mathbf{k}$, and a sampling of position vectors of points along the curve. (*Important:* This command and the next one require loading the `Vectors`` package.)

```
Show[Table[Vector[γ[t]], {t, 1.5, 2.5, .1}], ijk,
  ParametricPlot3D[γ[t], {t, 0, 2 π},
    PlotStyle → Thickness[.007]], ViewPoint → {6, 2, 2}]
```

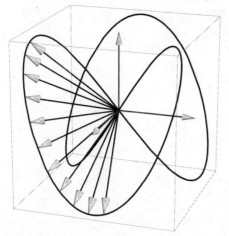

This `Manipulate` panel lets you trace out the curve with the tip of the position vector:

```
Manipulate[Show[Vector[γ[t]], ijk,
  ParametricPlot3D[γ[s], {s, 0, t}, PlotStyle → {Thick, Red}],
  PlotRange → {{-1, 1}, {-1, 1}, {-1, 1}},
  ViewPoint → {6, 2, 2}], {t, .01, 2 π}]
```

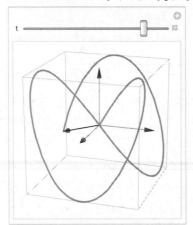

■ Intersecting Surfaces

The intersection of two surfaces is often a space curve. When the surfaces are the graphs of $z = f(x, y)$ and $z = g(x, y)$, it is sometimes practical to use cylindrical coordinates to obtain a parametrization of the intersection. This is illustrated in the following example.

■ Example 3.1.3

Find a parametrization for the curve of intersection of the paraboloid and plane given by

$$f(x, y) = 4 - x^2 - y^2 \text{ and } g(x, y) = 2 - x/5 + y/2.$$

Let's first define functions and get a nice picture of what we're after.

```
f[x_, y_] := 4 - x² - y²; g[x_, y_] := 2 - x / 5 + y / 2;
Show[surf1 = Plot3D[f[x, y], {x, -2, 2}, {y, -2, 2}],
  surf2 = Plot3D[g[x, y], {x, -2, 2}, {y, -2, 2}],
  ViewPoint → {6, 2, 3}, AxesLabel → {x, y, z},
  PlotRange → {0, 4}, BoxRatios → Automatic]
```

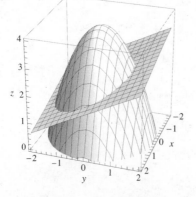

Certainly our plan will involve setting the two functions equal to each other. Doing so should yield a description of the projection of the curve onto the xy-plane. The key to obtaining a *useful* description of that curve is to work in cylindrical coordinates. So we'll substitute $x = r \cos t$ and $y = r \sin t$ into each of f and g. Then we'll set those expressions equal to each other and solve for r in terms of t. (Notice t is playing the role of the angle θ here.)

```
f[r Cos[t], r Sin[t]] == g[r Cos[t], r Sin[t]]
```

$$4 - r^2 \, \text{Cos}[t]^2 - r^2 \, \text{Sin}[t]^2 == 2 - \frac{1}{5} r \, \text{Cos}[t] + \frac{1}{2} r \, \text{Sin}[t]$$

```
roft = r /. First@Solve[%, r] // Simplify
```

$$\frac{1}{20} \left(2 \, \text{Cos}[t] - 5 \, \text{Sin}[t] - \sqrt{\left(804 \, \text{Cos}[t]^2 - 20 \, \text{Cos}[t] \, \text{Sin}[t] + 825 \, \text{Sin}[t]^2 \right)} \right)$$

With this done, parametric equations for the curve are

$$x = r(t) \cos t, \quad y = r(t) \sin t, \quad z = g\left(r(t) \cos t, r(t) \sin t\right).$$

```
z[t_] := g[roft Cos[t], roft Sin[t]];
γ[t_] = {roft Cos[t], roft Sin[t], z[t]} // Simplify
```

$$\left\{ \frac{1}{20} \, \text{Cos}[t] \left(2 \, \text{Cos}[t] - 5 \, \text{Sin}[t] - \sqrt{804 \, \text{Cos}[t]^2 - 20 \, \text{Cos}[t] \, \text{Sin}[t] + 825 \, \text{Sin}[t]^2} \right), \right.$$

$$-\frac{1}{20} \, \text{Sin}[t] \left(-2 \, \text{Cos}[t] + 5 \, \text{Sin}[t] + \sqrt{804 \, \text{Cos}[t]^2 - 20 \, \text{Cos}[t] \, \text{Sin}[t] + 825 \, \text{Sin}[t]^2} \right),$$

$$\frac{1}{400} \left(771 + 21 \, \text{Cos}[2 \, t] + 2 \, \text{Cos}[t] \, \sqrt{3258 - 42 \, \text{Cos}[2 \, t] - 40 \, \text{Sin}[2 \, t]} \right. -$$

$$\left. \left. 5 \, \text{Sin}[t] \, \sqrt{3258 - 42 \, \text{Cos}[2 \, t] - 40 \, \text{Sin}[2 \, t]} + 20 \, \text{Sin}[2 \, t] \right) \right\}$$

This is quite a mess—something we would *not* like to do "by hand." But now that *Mathematica* has done the algebra, we can now easily plot the curve.

```
curve = ParametricPlot3D[γ[t], {t, 0, 2 π},
    PlotStyle → AbsoluteThickness[2], ViewPoint → {6, 2, 3}]
```

Finally, we'll plot the curve of intersection along with each of the two surfaces.

```
GraphicsRow[{Show[surf1, curve, ViewPoint → {6, 2, 3},
    BoxRatios → Automatic, PlotRange → {0, 4}], Show[surf2,
    curve, ViewPoint → {6, 2, 3}, BoxRatios → Automatic]}]
```

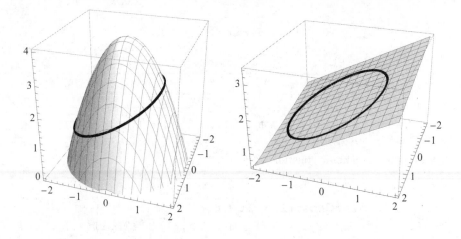

◆ Exercises

For each of the plane curves in Exercises 1-5, follow Example 3.1.1 to (a) plot the curve along with a sampling of position vectors, and (b) make a `Manipulate` panel that traces out the curve with the tip of the position vector.

1. $\gamma(t) = \langle \cos t, \sin t \rangle$

2. $\gamma(t) = \langle \cos(t/2), \sin t \rangle$

3. $\gamma(t) = \langle t, \cos t \rangle, \ 0 \le t \le 2\pi$

4. $\gamma(t) = \langle e^{-t/2} \cos t, e^{-t/2} \sin t \rangle, \ 0 \le t \le 2\pi$

5. $\gamma(t) = \langle 2\sin(t/5) + \sin t, 2\cos(t/5) + \cos t \rangle$

For each of the space curves in Exercises 6-8, follow Example 3.1.2 to (a) plot the curve along with a sampling of position vectors, and (b) make a `Manipulate` panel that traces out the curve with the tip of the position vector.

6. $\gamma(t) = \langle \cos t, \sin t, 2t^2/(3 + t^2) \rangle, \ 0 \le t \le 4\pi$

7. $\gamma(t) = \langle \cos(t/2), \sin t, t/6 \rangle, \ 0 \le t \le 2\pi$

8. $\gamma(t) = \langle e^{-t} \cos t, e^{-t} \sin t, t/6 \rangle, \ 0 \le t \le 4\pi$

For each of the pairs of surfaces in Exercises 9 and 10, follow Example 3.1.3 to find and plot a parametrization of the intersection.

9. $z = 4 - x^2 - y^2$ and $z = x^2 + 2y^2 - xy$

10. $z = 2x^2 + 2y^2 - 3$ and $z = x^2 - y^2$

3.2 Derivatives

Derivatives of vector functions are discussed in Section 14.2 of Stewart's *Calculus*. (ET:13.2)

The derivative of a vector-valued function is defined in a manner analogous to that of an ordinary real-valued function.

$$\gamma'(t) = \lim_{h \to 0} \frac{\gamma(t+h) - \gamma(t)}{h}$$

Notice that the difference quotient on the right side is a vector quantity; hence so is its limit as $h \to 0$. In fact,

$\gamma'(t_0)$ *is a vector that is tangent to the curve at* $\gamma(t_0)$, *provided that it exists and is not* **0**.

(This interpretation tells us about only the *direction* of $\gamma'(t_0)$; we will soon learn how to interpret its length.)

- ## Example 3.2.1

To illustrate the limiting process that defines $\gamma'(t_0)$, let's plot the curve

```
γ[t_] := {Sin[t / 2], 1 / 2 Sin[t]}
```

together with vectors given by the difference quotient

```
diffquot[t0_, h_] = Simplify[(γ[t0 + h] - γ[t0]) / h]
```

$$\left\{ \frac{-\mathrm{Sin}\left[\frac{t0}{2}\right] + \mathrm{Sin}\left[\frac{h+t0}{2}\right]}{h}, \frac{-\mathrm{Sin}[t0] + \mathrm{Sin}[h+t0]}{2\,h} \right\}$$

for small values of h and with $t_0 = \pi/2$ and initial point corresponding to the position vector

$$\gamma(\pi/2) = \left\langle \sin(\pi/4), \tfrac{1}{2}\sin(\pi/2) \right\rangle = \left\langle \sqrt{1/2}, 1/2 \right\rangle.$$

The following `Manipulate` panel lets you observe how the difference-quotient vector becomes tangent to the curve as $h \to 0$.

```
curve = ParametricPlot[γ[t], {t, 0, π}]; t0 = π / 2;
Manipulate[Show[curve, Vector[{γ[t0], diffquot[t0, h]}],
  Graphics[{Red, PointSize[.03], {Point[γ[t0]], Point[γ[t0 + h]]}}],
  PlotRange → {{0, 1.1}, {0, .55}}], {h, 1.25, .001}]
```

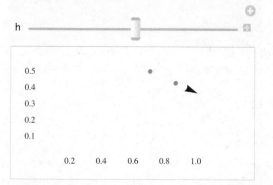

- **Example 3.2.2**

  Continuing with the same function (and `curve`) as in Example 3.2.1, let's plot a portion of the curve along with approximations to the derivative at several points along the curve. We'll compute the approximations by evaluating the difference quotient with $h = .0001$.

  ```
  Show[curve, Show[Table[Vector[{γ[t], diffquot[t, .0001]}],
     {t, 0, 4 π / 5, π / 10}]], PlotRange → All]
  ```

  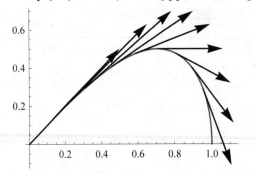

  With `Manipulate` you can move the (approximate) tangent vector along the curve.

  ```
  Manipulate[Show[curve, Vector[{γ[t], diffquot[t, .00001]}],
     Graphics[{Red, PointSize[.03], Point[γ[t]]}],
     PlotRange → {{0, 1.15}, {-.1, .72}}], {t, 0, 4 π / 5}]
  ```

  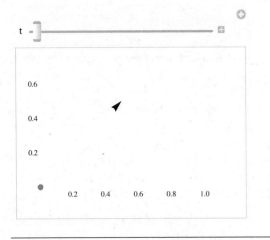

  Differentiation of vector-valued functions is done "component-wise" and is handled automatically by *Mathematica* (since the differentiation command has the `Listable` attribute):

  ```
  Clear[r, f, g, h];
  γ[t_] = {f[t], g[t], h[t]};
  γ'[t]
  D[γ[t], t]
  {f'[t], g'[t], h'[t]}
  {f'[t], g'[t], h'[t]}
  ```

▪ Example 3.2.3

Plot the space curve given by

```
γ[t_] := {Sin[t], Cos[t / 2], 1 / 2 + Sin[t / 2]}
```

along with the derivative as a tangent vector at each of $t = k\pi/4$, $k = 0, \ldots, 15$.

Let's first plot the curve and choose a good `ViewPoint`. Note that the curve will be completely drawn with $0 \le t \le 4\pi$. (Why?)

```
curve = ParametricPlot3D[γ[t], {t, 0, 4 π}, ViewPoint → {3, 4, 2}]
```

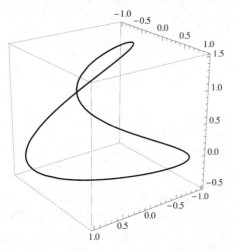

Now let's create a `Table` of vectors and plot them along the curve (without the bounding box and axes).

```
vecs = Table[Vector[{γ[t], γ'[t]}], {t, 0, 4 π, π / 4}];
Show[curve, vecs, Axes → None, PlotRange → All]
```

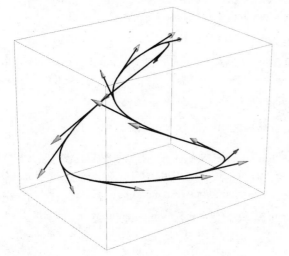

Here's an animation that shows the tangent vector moving along the curve.

```
Animate[Show[curve, Vector[{γ[t], γ'[t]}], Axes → None,
    PlotRange → {{-1.5, 1.5}, {-1.5, 1.5}, {-.7, 1.7}}],
    {t, 0, 4 π}]
```

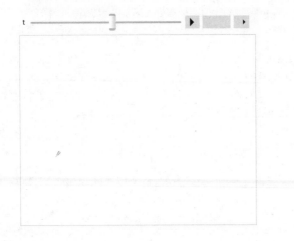

◆ Exercises

11-15. For each of the plane curves in Exercises 1-5 (after Section 3.1), (a) plot the curve along with a sampling of derivative vectors, and (b) make a `Animate` panel that show the derivative vector moving along the curve.

16-18. For each of the space curves in Exercises 6-8 (after Section 3.1), (a) plot the curve along with a sampling of derivative vectors, and (b) make a `Animate` panel that show the derivative vector moving along the curve.

19. Let C be the space curve parametrized by $\gamma(t) = 4\left\langle \sin^2 2t,\ \cos^2 3t,\ \sin t \right\rangle$.

   a) Find a value of t such that $\gamma(t) = \langle 3, 3, 1 \rangle$.

   b) Find a unit vector that is tangent to C at $(3, 3, 1)$.

   c) Find the equation of the plane that passes through $(3, 3, 1)$ and is orthogonal to C at that point.

20. Let C_γ and C_ρ be the two space curves parametrized respectively by

$$\gamma(t) = \left\langle \sin 2\pi t,\ \cos^2 \pi t,\ 2t \right\rangle \text{ and } \rho(t) = \langle 2t,\ 2 - 3t,\ 1 - t \rangle$$

   a) Find the point at which C_γ and C_ρ intersect. (*Hint:* The position vector of this point may be attained by $\gamma(t)$ and $\rho(t)$ at different values of t; so look at $\gamma(s) = \rho(t)$. The third coordinates give an easy relationship between s and t.)

   b) Find the (acute) angle between C_γ and C_ρ at the point where they intersect.

3.3 Parametrization

Parametrization and smoothness are discussed in Sections 14.2-14.3 of Stewart's *Calculus*.

A space curve C is merely a collection of points. It can be "traced out" by many different vector-valued functions. Vector-valued functions that trace out C are *parametrizations* of C.

- **Example 3.3.1**

Each of the vector functions

```
p[t_] := {Cos[t], Sin[t]};
q[t_] := {Sin[t² / (2 π)], Cos[t² / (2 π)]};
```

traces out the unit circle in the plane exactly once for $0 \le t \le 2\pi$. If we were to plot only the curve, then each parametrization would produce exactly the same picture. However, if we also plot several of the tangent vectors given by each derivative, we can see an indication of the different way that each parametrization traces out the curve.

```
Needs["Vectors`"];
opts := {PlotRange → {{-2, 2}, {-1.8, 2.1}}, Ticks → None};
GraphicsRow[{Show[ParametricPlot[p[t], {t, 0, 2 π}, Evaluate[opts]],
    Table[Vector[{p[t], p'[t]}], {t, π / 6, 2 π, π / 6}]],
  Show[ParametricPlot[q[t], {t, 0, 2 π}, Evaluate[opts]],
    Table[Vector[{q[t], q'[t]}], {t, π / 6, 2 π, π / 6}]]}]
```

Each of the two collections of tangent vectors shown above are plotted at the same set of equally spaced values of t. With the first parametrization, the initial points of the tangent vectors are also equally spaced and the vectors all have the same length. With the other parametrization, the spacing and the lengths of the vectors vary. Notice that shorter tangent vectors tend to be closer together along the curve.

Of course, an animation is much better at showing how the curve is traced out. Here's a function that makes an animation of the unit circle being traced out by a given parametrization $\gamma(t)$ for $t_1 \le t \le t_2$.

```
circleTangents[γ_Symbol, {t_, t1_, t2_}] := Animate[Show[
    Graphics[Circle[]], Vector[{γ[t], γ'[t]}],
    Graphics[{Red, PointSize[.025], Point[γ[t]]}], Axes → True,
    Ticks → None, PlotRange → 2 {{-1, 1}, {-1, 1}}], {t, t1, t2}]
```

Now we can animate the two parametrizations *p* and *q* side by side as follows. Note that the first argument to `circleTangents` is the *name* of the parametrization, not its formula.

```
Grid[
  {{circleTangents[p, {t, 0, 4 π}], circleTangents[q, {t, 0, 4 π}]}}]
```

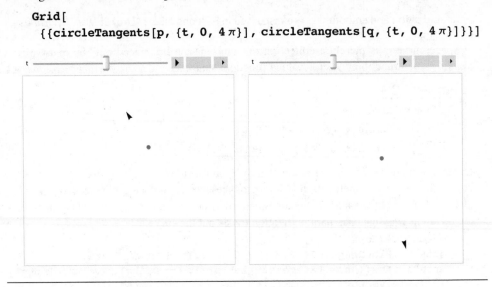

Often the parameter by which a curve is parametrized has a physical or geometric significance. Probably the most common and important situation is the one in which the parameter *t* represents *time*, and $\gamma(t)$ is the *position* (vector) of a moving object at time *t*. *Different parametrizations of the same path represent different ways in which an object could travel along that path*—with different speeds, different starting points, etc.

In other situations, a curve may be parametrized by some natural spatial coordinate or the angle formed by points on the curve relative to some fixed ray. The top half of the unit circle provides a simple illustration. We might decide to parametrize the curve by the counter-clockwise angle that position vectors form relative to the positive *x*-axis. This gives the familiar, standard parametrization

```
Clear[r]; γ[θ_] := {Cos[θ], Sin[θ]}
```

```
ParametricPlot[γ[θ], {θ, 0, π}]
```

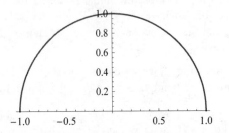

Similar parametrizations would be:

- $\mathbf{q}(\theta) = \langle -\cos\theta, \sin\theta \rangle$, which traces out the curve for $0 \le \theta \le \pi$, where θ is the clockwise angle that position vectors form relative to the negative *x*-axis;

$\mathbf{r}(\theta) = \langle -\sin\theta, \cos\theta \rangle$, which traces out the curve for $-\pi/2 \le \theta \le \pi/2$, where θ is the counter-clockwise angle that position vectors form relative to the positive x-axis.

We could also choose to let the x-coordinate be the parameter. Then the parametrization and the plot are obtained as follows. (The plot is not shown since it is identical to the one above.)

$$\gamma[x\_] := \left\{ x, \sqrt{1 - x^2} \right\}$$

ParametricPlot[γ[x], {x, -1, 1}, AspectRatio → Automatic]

In fact, any plane curve that is the graph of a function f has the simple parametrization $\gamma(x) = \langle x, f(x) \rangle$, where x varies over the domain of f.

■ Smoothness

We say that a vector-valued function $\gamma(t)$ on an interval $a \le t \le b$ is **smooth** if $\gamma'(t)$ is continuous and $\gamma'(t) \ne \mathbf{0}$ for all $a < t < b$. These two conditions prevent both discontinuities and reversal in the tangent vector's direction.

A space curve C is said to be **smooth** if it has a smooth parametrization. (This does not require every parametrization of the curve to be smooth.)

■ Example 3.3.2

Anyone would agree that the unit circle in the plane is a "smooth" curve, and according to our definition it is, because one parametrization of it is

γ[t_] := {Cos[t], Sin[t]}

for $0 \le t \le 2\pi$, which is a smooth vector-valued function, because its derivative

γ'[t]

{-Sin[t], Cos[t]}

is continuous and never $\mathbf{0}$. However, it turns out that

q[t_] := {Cos[π Sin[t / 2]], Sin[π Sin[t / 2]]}

is a different parametrization of the same curve. This is *not* a smooth vector-valued function, because of the $\cos(t/2)$ factor in both components of

q'[t]

$$\left\{ -\frac{1}{2}\pi \cos\left[\frac{t}{2}\right] \sin\left[\pi \sin\left[\frac{t}{2}\right]\right], \; \frac{1}{2}\pi \cos\left[\frac{t}{2}\right] \cos\left[\pi \sin\left[\frac{t}{2}\right]\right] \right\}$$

which causes $\mathbf{q}'(t)$ to be $\mathbf{0}$ when t is any odd multiple of π.

So, *not every parametrization of a smooth curve is smooth!* (However, $\mathbf{q}(t)$ restricted to the interval $-\pi \le t \le \pi$ *is* a smooth parametrization of the curve.)

It is interesting to animate the tangent vectors for this non-smooth parametrization of the unit circle. We'll use the `circleTangents` function from the preceding example. It will show the curve being traced traced out twice between $t = -\pi$ and $t = 3\pi$, an interval whose length is equal to the period of the parametrization. Also notice that the tangent vector reverses directionwhen t is an odd multiple of π.

```
circleTangents[q, {t, -π, 3 π}]
```

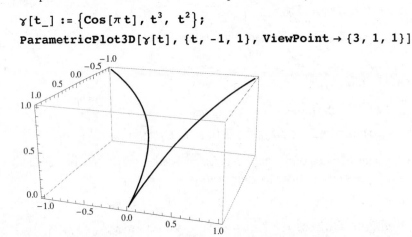

- ■ **Example 3.3.3**

An example of a *non-smooth curve* is the one parametrized by

```
γ[t_] := {Cos[π t], t³, t²};
ParametricPlot3D[γ[t], {t, -1, 1}, ViewPoint → {3, 1, 1}]
```

Notice that $\gamma'(0) = \mathbf{0}$. Moreover, it is not possible for this curve to have a smooth parametrization. However, it is *piecewise smooth*. The parametrization $\gamma(t)$ is smooth when restricted to either $(-\infty, 0]$ or $[0, \infty)$.

3.4 Arc Length and Curvature

Arc length and curvature are developed in Section 14.3 of Stewart's *Calculus*. (ET:13.3)

The length of a space curve with a smooth parametrization

$$\gamma(t) = \langle\, f(t),\, g(t),\, h(t)\,\rangle,\ \ a \le t \le b,$$

is the *integral of the length of the derivative* over the interval $a \le t \le b$:

$$L = \int_a^b \|\gamma'(t)\|\, dt = \int_a^b \sqrt{f'(t)^2 + g'(t)^2 + h'(t)^2}\ dt.$$

- ## Example 3.4.1

Plot the space curve described by

```
γ[t_] := {Cos[eᵗ], Sin[eᵗ], eᵗ/6}
```

for $0 \le t \le 2$ *and compute its length.*

It is easy to see that $\mathbf{r}(t)$ is just a peculiar parametrization of a simple helix with radius 1, and the graph indicates that this is indeed true.

```
ParametricPlot3D[γ[t], {t, 0, 2},
  ViewPoint → {6, 2, 2}, Axes → False]
```

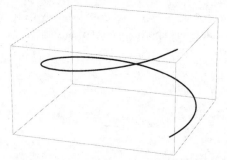

To compute the arc length, we integrate the length of the derivative over [0, 2]:

```
∫₀² √(γ'[t].γ'[t]) dt // Simplify
```

$$\frac{1}{6} \sqrt{37}\ \left(-1 + e^2\right)$$

A numerical approximation of this number is

```
% // N
```

```
6.47719
```

- ## Example 3.4.2

Plot the space curve described by

```
γ[t_] := {Sin[t], Cos[t], Sin[2 t]}
```

for $0 \le t \le \pi$ *and compute its length.*

This is a plot of the curve:

```
ParametricPlot3D[γ[t], {t, 0, π},
    Axes → False, ViewPoint → {6, 2, 2}]
```

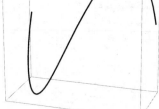

Here's its length:

$$\int_0^\pi \sqrt{\gamma'[t] \cdot \gamma'[t]} \, dt$$

$$2\sqrt{5} \ \text{EllipticE}\left[\frac{4}{5}\right]$$

Mathematica has given the answer in terms of an *elliptic integral*, one of many *special functions* that are built into *Mathematica*. A numerical approximation of this number is

```
% // N
```

```
5.27037
```

■ Example 3.4.3

Plot the space curve described by

```
γ[t_] := {Sin[π t], t², t}
```

for $0 \le t \le 2$ *and compute its length.*

```
ParametricPlot3D[γ[t], {t, 0, 2}, Axes → False]
```

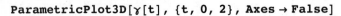

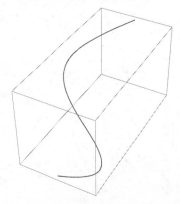

This is the length of the curve:

$$\int_0^2 \sqrt{\gamma'[t] \cdot \gamma'[t]} \ dt$$

$$\int_0^2 \sqrt{1 + 4\,t^2 + \pi^2 \, \text{Cos}[\pi\,t]^2} \ dt$$

Here *Mathematica* is at a loss as to what to do with the integral; most likely, it cannot be exactly evaluated by any means. However, a numerical approximation of the integral is

% // N

6.44158

■ The Arc-length Function

Given a space curve with a smooth parametrization

$$\gamma(t) = \langle f(t), g(t), h(t) \rangle$$

and a specified t_0, we can define a real-valued function $s(t)$ as the arc length

$$s(t) = \int_{t_0}^{t} \| \gamma'(u) \| \, du.$$

■ Example 3.4.4

Suppose that, for $t \geq 0$,

γ[t_] := {t², t³, t⁴}

Plot the curve and the graph of the arc length $s(t)$ for $0 \leq t \leq 1$.

The length of the derivative is

derLen[t_] = $\sqrt{\gamma'[t] \cdot \gamma'[t]}$

$$\sqrt{4\,t^2 + 9\,t^4 + 16\,t^6}$$

So the arc-length function corresponding to $t_0 = 0$ is

s[t_] = Assuming$\left[t > 0, \int_0^t \text{derLen[u]} \ du \right]$ // Simplify

$$\frac{1}{2048} \left(512\,t^2 \sqrt{4 + 9\,t^2 + 16\,t^4} \ + \right.$$

$$\left. 144 \left(-2 + \sqrt{4 + 9\,t^2 + 16\,t^4} \right) + 350\,\text{ArcSinh}\left[\frac{9 + 32\,t^2}{5\,\sqrt{7}} \right] - 175\,\text{Log}\left[\frac{25}{7} \right] \right)$$

The following shows the curve and the graph of $s(t)$ for $0 \le t \le 1$.

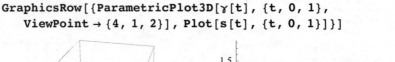

```
GraphicsRow[{ParametricPlot3D[γ[t], {t, 0, 1},
    ViewPoint → {4, 1, 2}], Plot[s[t], {t, 0, 1}]}]
```

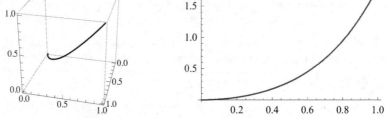

The reason the arc-length function is so important is that it provides a special parameter with which to parametrize a curve. If a given space curve is parametrized by

```
γ[s_] = {f[s], g[s], h[s]}

{f[s], g[s], h[s]}
```

where s is arc length measured from some fixed point on the curve, then the equation

$$\int_0^s \sqrt{\gamma'[u] \cdot \gamma'[u]} \ du == s$$

$$\int_0^s \sqrt{f'[u]^2 + g'[u]^2 + h'[u]^2} \ du == s$$

is true. Now differentiating each side of this equation with respect to s gives

```
D[%, s]
```

$$\sqrt{f'[s]^2 + g'[s]^2 + h'[s]^2} == 1$$

This shows that:

💧   *the derivative of a curve parametrized by arc length is always a unit vector.*

Parametrization of a smooth curve by arc length is always theoretically possible, but it is practical (in an explicit way) in only a few simple cases.

A simple example of a curve parametrized by arc length is the unit circle parametrized by

```
γ[t_] := {Cos[t], Sin[t]}
```

Here we usually think of t as the radian measure of the angle that the position vector makes with the x-axis, but since the radius of the circle is 1, this corresponds exactly to the arc length subtended by the angle. Note that the derivative has constant length 1:

```
γ'[t]
√γ'[t].γ'[t]  // Simplify

{-Sin[t], Cos[t]}

1
```

■ Curvature

Suppose that C is a curve with smooth parametrization $\gamma(t)$ (i.e., γ is differentiable with $\gamma'(t) \neq \mathbf{0}$). Then $\gamma'(t)$ provides us with a tangent vector to the curve at any given point, but the length of that vector is dependent on the parametrization. However, at each point along the curve there is a **unit tangent vector T** that is characteristic of the curve itself and not dependent on the parametrization. The unit tangent vector is parametrized by

$$\mathbf{T}(t) = \frac{\gamma'(t)}{\| \gamma'(t) \|} .$$

A *Mathematica* definition of the unit tangent vector is

```
unitTan[γ_, t_Symbol] := γ'[t] / √( γ'[t].γ'[t] )
```

Note that the first argument to `unitTan` is the *name* of the parametrization function γ, not the expression $\gamma(t)$.

■ Example 3.4.5

a) *Plot the curve parametrized by*

```
γ[t_] := {2 Cos[t], 2 Sin[t], Sin[3 t]}
```

for $0 \le t \le \pi$ along with several tangent vectors given by $\gamma'(t)$.

b) *Plot the same curve along with several unit tangent vectors.*

Let's do both parts at once and plot the results side by side. Here's the unit tangent vector:

```
T[t_] = unitTan[γ, t]
```

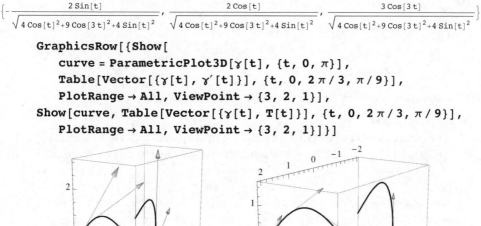

```
GraphicsRow[{Show[
    curve = ParametricPlot3D[γ[t], {t, 0, π}],
    Table[Vector[{γ[t], γ'[t]}], {t, 0, 2 π / 3, π / 9}],
    PlotRange → All, ViewPoint → {3, 2, 1}],
  Show[curve, Table[Vector[{γ[t], T[t]}], {t, 0, 2 π / 3, π / 9}],
    PlotRange → All, ViewPoint → {3, 2, 1}]}]
```

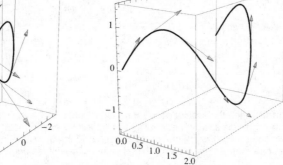

Since the unit tangent vector $\mathbf{T}(t)$ has constant length, only its direction changes as t varies. The derivative of $\mathbf{T}$ with respect to arc length provides a *parametrization-independent* indication of the way in which the direction of the curve is changing at any given point on the curve. In fact, the **curvature** of a curve is defined by

$$\kappa = \left| \frac{d\mathbf{T}}{ds} \right|.$$

Application of the Chain Rule (see Section 10.3 of Stewart) provides a more convenient formula for κ that does not require parametrization by arc length. Given any smooth parametrization, the curvature at the point whose position vector is $\mathbf{r}(t)$ is given by

$$\kappa(t) = \frac{\left| \mathbf{T}'(t) \right|}{\left| \gamma'(t) \right|}.$$

So let's make the following *Mathematica* definition.

```
kappa[γ_, t_Symbol] := √( (D[unitTan[γ, t], t].D[unitTan[γ, t], t]) / (γ'[t].γ'[t]) )
```

- **Example 3.4.6**

Compute the curvature of the ellipse parametrized by

```
el[t_] := {3 Cos[t], Sin[t]}
```

Plot the graph of the curvature for $0 \le t \le 2\pi$ and interpret its behavior.

The curvature is

```
κ[t_] = kappa[el, t] // Simplify
```

$$3 \sqrt{-\frac{1}{(-5+4 \operatorname{Cos}[2\,t])^3}}$$

The following `GraphicsRow` shows the ellipse and its curvature.

```
GraphicsRow[
  {ParametricPlot[el[t], {t, 0, 2 π}, AspectRatio → Automatic],
   Plot[κ[t], {t, 0, 2 π}, AspectRatio → Automatic]}, Spacings → 0]
```

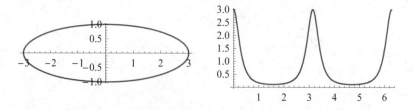

Interpretation: Notice that peaks occur at multiples of π. These correspond to points where the ellipse crosses the x-axis, which is where the ellipse curves the most—that is, changes direction most rapidly. The curvature is a minimum at each odd multiple of $\pi/2$. These correspond to points where the ellipse crosses the y-axis, which is where the ellipse curves the least.

■ Example 3.4.7

Compute the curvature of the curve parametrized by

```
r[t_] := {2 Cos[t], 2 Sin[t], Sin[5 t]}
```

Plot the graph of the curvature for $0 \le t \le 2\pi$ and interpret its behavior.

The curvature is

```
κ[t_] = kappa[r, t] // Simplify
```

$$4\sqrt{\frac{658 - 600\,\text{Cos}[10\,t]}{(33 + 25\,\text{Cos}[10\,t])^3}}$$

The following `GraphicsRow` shows the curve and its curvature.

```
GraphicsRow[
  {ParametricPlot3D[r[t], {t, 0, 2 π}, AspectRatio → Automatic],
   Plot[κ[t], {t, 0, 2 π}]}]
```

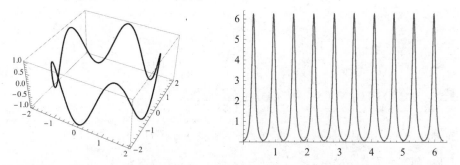

Interpretation: Notice that the graph of the curvature has ten peaks. Each peak corresponds to a point on the space curve where the *z*-coordinate is a maximum or minimum, since these are the points where the space curve changes direction most rapidly. The curvature is very close to zero at values of *t* corresponding to points where the space curve is nearly straight.

■ Example 3.4.8

Determine, as a function of x, the curvature of a plane curve with equation $y = f(x)$. Then apply the result to find and graph the curvature of $y = x^3$.

First we need to parametrize the curve. Using *x* as the parameter, we have

```
Clear[f]; γ[x_] := {x, f[x]}
```

Then the curvature is given by

```
kappa[γ, x] // Simplify
```

$$\sqrt{\frac{f''[x]^2}{\left(1 + f'[x]^2\right)^3}}$$

Now if

```
f[x_] := x^3
```

then the curvature is given by

$$\kappa[x\_] = \text{kappa}[\gamma, x] \text{ // Simplify}$$

$$6 \sqrt{\frac{x^2}{\left(1+9\,x^4\right)^3}}$$

The graphs of the curve and its curvature are shown below.

$$\text{Plot}[\{f[x], \kappa[x]\}, \{x, -1.5, 1.5\}]$$

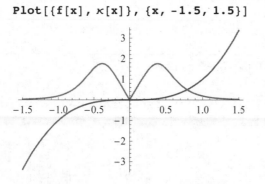

3.5 Velocity and Acceleration

Section 14.4 in Stewart's *Calculus* deals with velocity and acceleration. (ET:13.4)

Probably the most interesting and important application of parametrized curves is in the study of the motion of a moving object. If we interpret a vector-valued function **r**(*t*) as the position vector of an object at time *t*, then the **velocity** of the object at time *t* is a vector quantity and is in fact the derivative of **r**:

$$\mathbf{v}(t) = \mathbf{r}'(t).$$

The length of the velocity vector is the **speed** of the object. As is to be expected, the derivative of velocity is **acceleration**, again a vector quantity:

$$\mathbf{a}(t) = \mathbf{v}'(t) = \mathbf{r}''(t).$$

■ Example 3.5.1

Suppose that the position vector of an object moving in the plane is

$$\text{r}[t\_] := \{2\,\text{Cos}[t\,/\,6] + \text{Cos}[5\,t\,/\,6], \ 2\,\text{Sin}[t\,/\,6] + \text{Sin}[5\,t\,/\,6]\}$$

Compute the velocity and acceleration as functions of t. Then plot the curve for $0 \le t \le 4\pi$ along with velocity and acceleration vectors on that interval at multiples of $\pi/3$. Comment on the relationship between the change in the direction of the object's path and the direction of the acceleration vector. Finally, compute the speed as a function of t on the same interval.

First we calculate the velocity and acceleration vectors.

```
v[t_] = r'[t]
```

$$\left\{-\frac{1}{3}\sin\left[\frac{t}{6}\right]-\frac{5}{6}\sin\left[\frac{5t}{6}\right],\ \frac{1}{3}\cos\left[\frac{t}{6}\right]+\frac{5}{6}\cos\left[\frac{5t}{6}\right]\right\}$$

```
a[t_] = r''[t]
```

$$\left\{-\frac{1}{18}\cos\left[\frac{t}{6}\right]-\frac{25}{36}\cos\left[\frac{5t}{6}\right],\ -\frac{1}{18}\sin\left[\frac{t}{6}\right]-\frac{25}{36}\sin\left[\frac{5t}{6}\right]\right\}$$

The following displays the curve and the velocity and acceleration vectors.

```
Show[ParametricPlot[r[t], {t, 0, 4 π}],
 Table[Vector[{r[t], v[t]}], {t, 0, 4 π, π / 3}],
 Table[Vector[{r[t], a[t]}], {t, 0, 4 π, π / 3}],
 Axes → False, PlotRange → All]
```

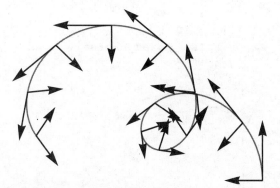

Comment: The path of the object always bends in the direction of the acceleration vector. In this example, that's always a bend to the left. The acceleration vectors appear sometimes to be orthogonal to the path, but **not always.**

The speed of the object is:

```
√r'[t].r'[t]  // Simplify
Plot[%, {t, 0, 4 π}, PlotRange → {0, 1.5}]
```

$$\frac{1}{6}\sqrt{29+20\cos\left[\frac{2t}{3}\right]}$$

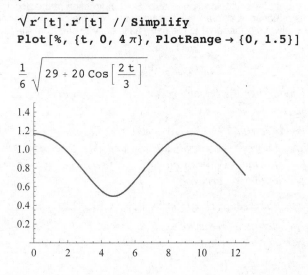

Example 3.5.2

Suppose that the position vector of an object moving in space is, for $t \geq 0$,

```
r[t_] := {Sin[t], Cos[t], t / (2 π)}
```

Compute the velocity and acceleration as functions of t. Then plot the curve for $0 \leq t \leq 2\pi$ along with velocity and acceleration vectors on that interval at multiples of $\pi/3$. What do you observe about the direction of the acceleration vector?

```
r'[t]
r"[t]
```

$$\left\{\text{Cos}[t], -\text{Sin}[t], \frac{1}{2\pi}\right\}$$

```
{-Sin[t], -Cos[t], 0}
```

```
Show[ParametricPlot3D[r[t], {t, 0, 2 π}, ViewPoint → {6, 2, 3}],
   Table[Vector[{r[t], r'[t]}], {t, 0, 2 π, π / 3}],
   Table[Vector[{r[t], r"[t]}], {t, 0, 2 π, π / 3}],
   ijk, PlotRange → All, Axes → False, Boxed → False]
```

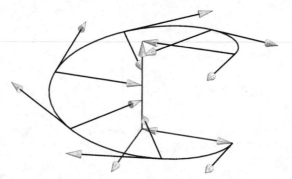

Observation: The vertical component of the acceleration is zero. So the direction of the acceleration vector is always horizontal. Its direction is also always toward the *z*-axis, the center of the helix. The acceleration vectors are always orthogonal to the path, as is easy to see from the fact that $\mathbf{r}'(t) \cdot \mathbf{r}''(t) = 0$.

■ Example 3.5.3

a) *Show that if $\mathbf{r}(t)$ has constant length, then the velocity vector $\mathbf{v}(t)$ is always orthogonal to the position vector $\mathbf{r}(t)$.*

b) *Show that if an object moves with constant (non-zero) speed, then its acceleration vector $\mathbf{a}(t)$ is orthogonal to its velocity vector $\mathbf{v}(t)$ whenever $\mathbf{a}(t) \neq \mathbf{0}$ and therefore orthogonal to the path of the object. What does this fact say about the forces experienced by a passenger in a car travelling at constant speed along a curvy road?*

c) *Verify that an object whose position vector is $\mathbf{r}(t) = \left\langle \sqrt{8t^3}, \sqrt{(9-2t)^3} \right\rangle$ travels with constant speed and that its acceleration is orthogonal to its path. Plot the path for $0 \leq t \leq 9/2$ along with several acceleration vectors.*

(a) Suppose that

r[t_] = {f[t], g[t], h[t]}

{f[t], g[t], h[t]}

If **r**(*t*) has constant length, then **r**(*t*) · **r**(*t*) is constant; therefore its derivative must be 0.

r[t].r[t] == k

D[%, t]

f[t]^2 + g[t]^2 + h[t]^2 == k

2 f[t] f'[t] + 2 g[t] g'[t] + 2 h[t] h'[t] == 0

The derivative of **r**(*t*) · **r**(*t*) is in fact 2 **r**(*t*) · **r**'(*t*) by the Chain Rule, or by observation of the calculation above. Therefore **r**(*t*) · **r**'(*t*) = 0 —that is, **r**(*t*) · **v**(*t*) = 0 —for all *t*.

(b) If an object moves with constant speed, then its velocity vector has constant length and is therefore always orthogonal to its derivative by the preceding argument. If a car goes around a curve at constant speed, its acceleration will be orthogonal to the road and in the direction (right or left) of the turn. The force experienced by a passenger is in the opposite direction, still orthogonal to the direction in which the car is moving—that is, the force is always purely "side-to-side" with no forward or backward component.

(c) The velocity vector and speed for

$$r[t\_] := \left\{ \sqrt{8\,t^3}\,,\ \sqrt{(9-2\,t)^3} \right\}$$

are respectively

v[t_] = r'[t]

$$\left\{ \frac{3\sqrt{2}\ t^2}{\sqrt{t^3}}\,,\ -\frac{3\,(9-2\,t)^2}{\sqrt{(9-2\,t)^3}} \right\}$$

$\sqrt{\text{v[t].v[t]}}$ **// Simplify**

9

So we see that the speed is always 9 (with appropriate units). The acceleration vector is

a[t_] = v'[t] // Simplify

$$\left\{ \frac{3\,t}{\sqrt{2}\ \sqrt{t^3}}\,,\ \frac{27-6\,t}{\sqrt{(9-2\,t)^3}} \right\}$$

The dot product of the velocity and acceleration shows that they are orthogonal:

v[t].a[t] // Simplify

0

The following is a plot of the curve and several acceleration vectors.

```
Show[ParametricPlot[r[t], {t, 0, 9/2}],
  Table[Vector[{r[t], a[t]}], {t, .02, 4.48, 4.46/20}],
  PlotRange → All]
```

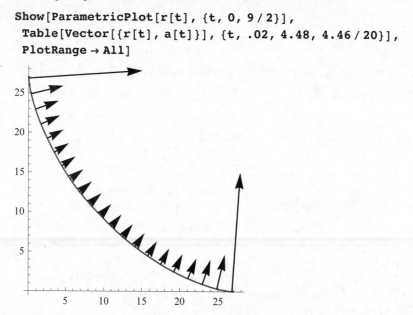

◆ Exercises

21-28. For each of the curves described in Exercises 1-8 (after Section 3.1), plot the curve and compute its arc length (exactly if possible).

29, 30. For the curves described in Exercises 3 and 8 (after Section 3.1), plot the parametrization's arc length function.

31-38. For each of the curves described in Exercises 1-8 (after Section 3.1), find the curvature and plot its graph.

39. Find the point(s) at which the graph of $y = x^4$ has maximum curvature.

40. Find the equation of the circle that best approximates the parabola $y = x^2$ at $(1/2, 1/4)$ in the sense that it is tangent to and has the same curvature as the parabola at that point. Create a plot of both the parabola and the circle. (Note: The curvature of a circle with radius R is $1/R$.)

41-48. For each of the curves described in Exercises 1-8 (after Section 3.1), plot the curve along with a sampling of unit tangent vectors.

49-56. For each of the curves described in Exercises 1-8 (after Section 3.1), plot the curve along with a sampling of acceleration vectors.

57. There are two definitions that one might give for a scalar acceleration of an object moving according to a parametrization $\mathbf{r}(t)$. One is $|\mathbf{r}''(t)|$, the length of the derivative of the velocity vector. The other is the derivative of the speed. Give an example that shows that these definitions, in general, produce different results.

3.6 Parametric Surfaces:
Vector-valued Functions of Two Variables

Parametric surfaces are the subject of Section 17.6 of Stewart's *Calculus*. (ET:16.6)

We have seen that space curves have parametrizations that take the form of vector-valued functions of one variable. Analogous parametrizations of *surfaces* take the form of vector-valued functions of *two* variables (i.e., two parameters).

■ Example 3.6.1

One parametrization of the unit sphere is

```
r[θ_, φ_] := {Cos[θ] Sin[φ], Sin[θ] Sin[φ], Cos[φ]}
```

The parameters θ and ϕ are the familiar spherical-coordinate angles. Note that the length of $\mathbf{r}(\theta, \phi)$ is always 1:

$$\sqrt{\mathbf{r}[\theta, \phi].\mathbf{r}[\theta, \phi]}$$

```
% // Simplify
```

$$\sqrt{\text{Cos}[\phi]^2 + \text{Cos}[\theta]^2 \text{Sin}[\phi]^2 + \text{Sin}[\theta]^2 \text{Sin}[\phi]^2}$$

1

Mathematica's `ParametricPlot3D` plots parametric surfaces as well as parametric curves in three dimensions. Note that the entire sphere is obtained with $0 \leq \theta \leq 2\pi$ and $0 \leq \phi \leq \pi$.

```
ParametricPlot3D[r[θ, φ], {θ, 0, 2 π}, {φ, 0, π}]
```

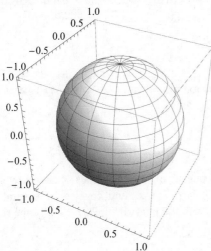

Example 3.6.2.

The following function is a parametrization of a *helical tube*.

```
r[t_, θ_] := 2 {Cos[t] (4 + Sin[θ]), Sin[t] (4 + Sin[θ]), t + Cos[θ]}
```

```
ParametricPlot3D[r[t, θ], {t, 0, 4 π}, {θ, 0, 2 π},
  ViewPoint → {3, 1, 1} ]
```

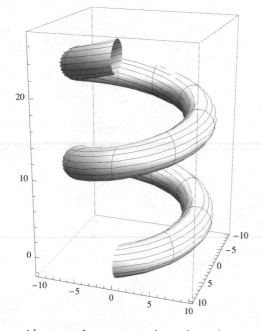

The *grid curves* that one sees in such a plot are actually parametric curves obtained by setting one of the parameters equal to a constant. For example, if we set $t = \pi/2$, we have the following function of one variable, which is easily seen to be a parametrization of a circle of radius 2 in the *yz*-plane.

```
r[π / 2, θ] // Expand
```

```
{0, 8 + 2 Sin[θ], π + 2 Cos[θ]}
```

If we set $\theta = \pi/2$, we have the following function of one variable, which is easily seen to be a parametrization of a helix centered along the *z*-axis.

```
r[t, π / 2]
```

```
{10 Cos[t], 10 Sin[t], 2 t}
```

The following creates three plots. The first consists of several circles obtained by setting t equal to multiples of $\pi/8$ in the interval $[0, 4\pi]$. The second consists of several helices obtained by setting θ equal to multiples of $\pi/4$ in the interval $[0, 2\pi]$. Finally the third plot simply superimposes the first two.

```
GraphicsRow[{Show[
  Table[ParametricPlot3D[r[t, θ], {θ, 0, 2 π}], {t, 0, 4 π, π / 6}],
  Ticks → None, PlotRange → All, ViewPoint → {3, 1, 1} ], Show[
  Table[ParametricPlot3D[r[t, θ], {t, 0, 4 π} ], {θ, 0, 2 π, π / 4}],
  Ticks → None, PlotRange → All, ViewPoint → {3, 1, 1} ]}]
```

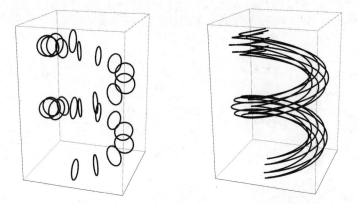

▪ Example 3.6.3

The following is an interesting surface that resembles a conch shell.

```
conch[u_, t_] :=
  {Cos[2 π u] u (1 + Sin[t]), Sin[2 π u] u (1 + Sin[t]), 4 - u (3 + Cos[t])}

ParametricPlot3D[conch[u, t], {u, 0, 2}, {t, 0, 2 π},
  PlotPoints → {30, 15}, ViewPoint → {3, 1, 1} , PlotRange → All]
```

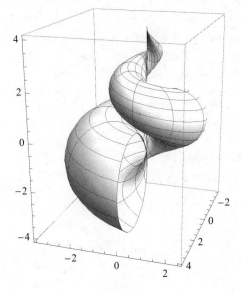

■ **Example 3.6.4**

This is known as *Dini's Surface*.

```
dini[u_, v_] :=
  1.5 {Cos[u] Sin[v], Sin[u] Sin[v], Cos[v] + Log[Tan[.5 v]] + .1 u}

ParametricPlot3D[dini[u, v], {u, 0, 5 Pi}, {v, .01, 2},
  PlotPoints → {40, 20}, Boxed → False, Axes → None]
```

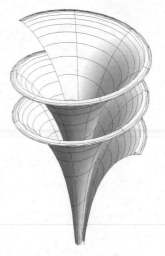

■ **The Normal Vector**

Given a surface with parametrization

$$\mathbf{r}(u, v) = \langle x(u, v), y(u, v), z(u, v) \rangle,$$

the partial derivatives $\mathbf{r}_u$ and $\mathbf{r}_v$ (see Section 4.2) are a pair of non-parallel tangent vectors to the surface at any given point, provide neither is **0**. As a result, their cross product $\mathbf{r}_u \times \mathbf{r}_v$ is a vector that is normal to the surface, and a **unit normal vector** is

$$\mathbf{n} = \frac{\mathbf{r}_u \times \mathbf{r}_v}{\| \mathbf{r}_u \times \mathbf{r}_v \|}.$$

■ **Example 3.6.5**

Consider the parametrization of the sphere from Example 3.6.1:

```
r[θ_, φ_] := {Cos[θ] Sin[φ], Sin[θ] Sin[φ], Cos[φ]}
```

We get tangent vectors by differentiating $\mathbf{r}(\theta, \phi)$ with respect to θ and with respect to ϕ:

```
{rθ[θ_, φ_] = ∂θ r[θ, φ], rφ[θ_, φ_] = ∂φ r[θ, φ]} // Simplify
```

{{-Sin[θ]Sin[φ], Cos[θ]Sin[φ], 0}, {Cos[θ]Cos[φ], Cos[φ]Sin[θ], -Sin[φ]}}

Their cross product gives a normal vector:

```
nrml = rθ[θ, φ] × rφ[θ, φ] // Simplify
```

{-Cos[θ] Sin[φ]², -Sin[θ] Sin[φ]², -Cos[φ] Sin[φ]}

The length of the normal vector is

```
√nrml.nrml // Simplify
```

$$\sqrt{\text{Sin}[\phi]^2}$$

which is simply $\sin\phi$, since $0 \le \phi \le \pi$. So a unit normal vector is

```
unitnrml[θ_, φ_] = nrml/Sin[φ] // Simplify
```

```
{-Cos[θ] Sin[φ], -Sin[θ] Sin[φ], -Cos[φ]}
```

Notice that this is just the negative of the position vector $\mathbf{r}(\theta, \phi)$; thus, when located at the corresponding point on the sphere, it points directly at the origin—as we might expect. Here's a plot showing these three vectors at the point corresponding to $\theta = \phi = \pi/3$.

```
r0=r[π/3,π/3]; sphr=
ParametricPlot3D[r[θ,φ], {θ,0,2 π}, {φ,0,π}, PlotStyle→Opacity[.25]];
Show[sphr, Vector/@{{r0,unitnrml[π/3,π/3]}, {r0,rθ[π/3,π/3]}, {r0,rφ[π/3,π/3]}},
ViewPoint → {6, 2, 2}, PlotRange → All]
```

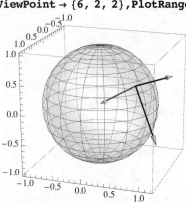

For an outward-pointing normal vector, we simply negate the previous one, obtaining precisely the position vector of the point. The following plot shows the outward-pointing normal and the position vector corresponding to $\theta = \phi = \pi/3$.

```
Show[Vector/@{r0, {r0,-unitnrml[π/3,π/3]}, {r0,rθ[π/3,π/3]}, {r0,rφ[π/3,π/3]}},
sphr, ViewPoint → {6, 2, 2}, PlotRange → All]
```

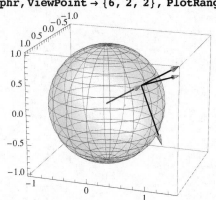

◆ **Exercises**

58. Following Example 3.6.2, plot grid curves for each of the surfaces in Examples 3.6.3 and 3.6.4.

59. For the torus parametrized by

$$\mathbf{r}(\theta, u) = \langle (2 + \cos u) \cos \theta, (2 + \cos u) \sin \theta, \sin u \rangle,$$

a) compute the normal vector $\mathbf{r}_\theta \times \mathbf{r}_u$,

b) make a plot showing the torus and the outward-pointing unit normal vector at the point corresponding to $\theta = \pi/3$ and $u = \pi/4$.

60. Given a plane curve parametrized by

$$\mathbf{q}(t) = \langle f(t), g(t) \rangle,$$

a parametrization of the corresponding surface of revolution about the z-axis is given by

$$Q(\theta, t) = \langle f(t) \cos \theta, f(t) \sin \theta, g(t) \rangle.$$

Plot the plane curve and the surface of revolution corresponding to:

a) $\mathbf{q}(t) = \langle 2 + \cos t, \sin t \rangle, \ 0 \le t \le 2\pi$

b) $\mathbf{q}(t) = \langle t, t^2 \rangle, \ 0 \le t \le 1$

c) $\mathbf{q}(t) = \langle 2 \sin t + \sin 5t, 2 \cos t + \cos 5t \rangle, \ -\pi/2 \le t \le \pi/2$

In each part, plot for $\pi/3 \le \theta \le 2\pi$ with `ViewPoint→{6,2,2}` or for $0 \le \theta \le 3\pi/2$ with the default `ViewPoint`.

4 Multivariate Functions

In this chapter we will explore *multivariate functions*: real-valued functions of two or more variables. These are functions that associate real numbers with points (or vectors) in two-, three-, or higher-dimensional space. Here we will use *Mathematica* to explore familiar differential calculus concepts in the context of multivariate functions: limits, continuity, differentiation, and optimization.

We have already encountered such functions in our brief investigation of surfaces in Chapter 2. We saw that the graph of a function of two variables is typically a surface in three dimensional space, as is a contour (or level surface) of a function of three variables. You should review Chapter 2 before proceeding further here.

> Section 15.1 of Stewart's *Calculus* discusses multivariate functions and related notions, including surfaces and contours. (ET:14.1)

4.1 Limits and Continuity

> The notions of limit and continuity for multivariate functions are developed in Section 15.2 of Stewart's *Calculus*. (ET:14.2)

Recall that a real-valued function of one variable, say $f(x)$, is continuous at a number a inside of an open interval contained in the domain of f, if $\lim_{x \to a} f(x)$ exists and equals $f(a)$. Continuity of a function of two variables, say $f(x, y)$, at a point (a, b) inside of an open *disk* contained in the domain of f, is defined essentially the same way:

☐ *f is continuous at a point (a, b) in its domain if* $\lim_{(x, y) \to (a, b)} f(x, y)$ *exists and equals* $f(a, b)$.

For functions of more than two variables, the definition is again basically the same: *limit = value*. So the existence of the limit is always at the heart of any continuity question. See Stewart's *Calculus* for a thorough discussion of limits.

The domain of a function of two variables, $f(x, y)$, is some subset of the xy-plane. Suppose that the point (a, b) lies in an open disk contained in the domain of f. In order for $\lim_{(x, y) \to (a, b)} f(x, y)$ to exist, $f(x, y)$ must approach the same limiting value as $(x, y) \to (a, b)$ along *every possible path* leading to (a, b). Consequently, if $f(x, y)$ approaches different limiting values as (x, y) approaches (a, b) along two different paths, then $\lim_{(x, y) \to (a, b)} f(x, y)$ does not exist. A similar statement is true for functions of three (or more) variables, and even for functions of one variable. (What is the analogous fact about functions of one variable?)

The following examples will illustrate this but mainly serve the purpose of indicating how very peculiarly functions of more than one variable can behave.

■ **Example 4.1.1**

Consider the function

$$f[x\_, y\_] := (y - x) \Big/ \sqrt{x^2 + y^2}$$

which is defined and continuous at all points in the *xy*-plane except $(0, 0)$, where it is undefined. The following shows the graph for $-3 \le x \le 3$ and $-3 \le y \le 3$.

```
Plot3D[f[x, y], {x, -3, 3}, {y, -3, 3}, MaxRecursion → 5]
```

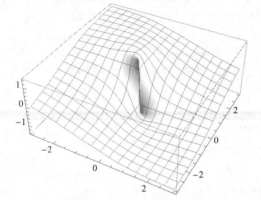

This function is not continuous at $(0, 0)$ simply because it is undefined there. The real question, then, is whether it might be possible to *make it continuous* by defining it appropriately at $(0, 0)$. Answering this question comes down to determining whether or not $\lim_{(x, y) \to (0, 0)} f(x, y)$ exists.

The trace in $x = 0$ is the graph of $f(x, y)$ along the *y*-axis. This trace is the graph of

$$z = f(0, y) = y \Big/ \sqrt{y^2} = y/|y|$$

and should be familiar to you:

```
Plot[f[0, y], {y, -3, 3}]
```

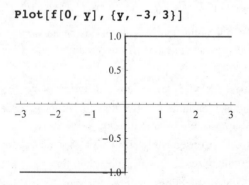

The graph shows that $f(x, y)$ approaches -1 as $(x, y) \to (0, 0)$ along the negative *y*-axis, while $f(x, y)$ approaches 1 as $(x, y) \to (0, 0)$ along the positive *y*-axis. So, the limit as $(x, y) \to (0, 0)$ does not exist. Therefore, defining f at $(0, 0)$ could never make f continuous there.

It is interesting to note that the trace in $y = x$ is the graph of $f(x, x) = 0$ (for $x \neq 0$). So $f(x, y)$ approaches 0 from either direction as $(x, y) \to (0, 0)$ along the line $y = x$. This is the only straight line through $(0, 0)$ along which $f(x, y)$ has a limit, as can be seen by looking at $f(x, y)$ along $y = mx$.

```
f[x, m x] // Simplify
```

$$\frac{(-1 + m)\, x}{\sqrt{\left(1 + m^2\right) x^2}}$$

Notice that $f(x, mx)$ is always a non-zero multiple of $x / |x|$ unless $m = 1$.

■ Example 4.1.2

Consider the function

```
f[x_, y_] := -x / (x² + y²)
```

which, just as the function in the last example, is defined and continuous at all points in the xy-plane except $(0, 0)$, where it is undefined. The following shows the graph for $-3 \leq x \leq 3$ and $-3 \leq y \leq 3$.

```
Plot3D[f[x, y], {x, -3, 3}, {y, -3, 3}, MaxRecursion → 5,
  BoxRatios → {1, 1, 1}, PlotRange → {-5, 5},
  ViewPoint → {6, 4, 2}, ClippingStyle → None]
```

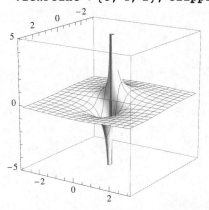

Clearly $\lim_{(x, y) \to (0, 0)} f(x, y)$ does not exist; in fact, any trace along a straight line $y = mx$ is the graph of

```
f[x, m x] // Factor
```

$$-\frac{1}{\left(1 + m^2\right) x}$$

and so has a vertical asymptote at $x = 0$. What is really interesting here is that *along certain paths*, $f(x, y)$ actually *has* a limit as $(x, y) \to (0, 0)$. Along the parabola $x = -y^2$, we have

```
f[-y², y] // Simplify
```

$$\frac{1}{1 + y^2}$$

whose graph is very simple:

$$\text{Plot}\left[\text{f}\left[-\text{y}^2,\ \text{y}\right],\ \{\text{y, -3, 3}\}\right]$$

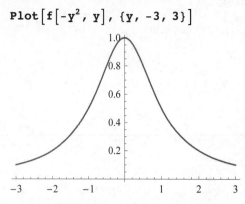

So we see that $f(x, y)$ approaches 1 as $(x, y) \to (0, 0)$ along the parabola $x = -y^2$. (In fact, $f(x, y)$ approaches k as $(x, y) \to (0, 0)$ along the parabola $x = -ky^2$.) Along the cubic curve $x = -y^3$, $f(x, y)$ approaches 0 as $(x, y) \to (0, 0)$:

$$\text{f}\left[-\text{y}^3,\ \text{y}\right]\ \text{// Simplify}$$

$$\frac{\text{y}}{1+\text{y}^4}$$

$$\text{Plot}\left[\text{f}\left[-\text{y}^3,\ \text{y}\right],\ \{\text{y, -3, 3}\}\right]$$

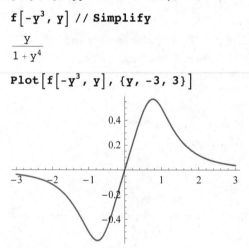

▪ Example 4.1.3

Consider yet another function

$$\text{f}[\text{x\_, y\_}] := \frac{\text{x y}}{\text{x}^2 + \text{y}^2}$$

that is defined and continuous at all points in the xy-plane except $(0, 0)$, where it is undefined. The following shows two views of the graph for $-1 \le x \le 1$ and $-1 \le y \le 1$. In order to get better resolution at the discontinuity, we'll set $\text{MaxRecursion} \to 5$

```
GraphicsRow[{surf1 = Plot3D[f[x, y], {x, -1, 1}, {y, -1, 1},
    Mesh → True, BoxRatios → Automatic, MaxRecursion → 5],
    Show[surf1, ViewPoint → {2.4, 1.3, 2}]}]
```

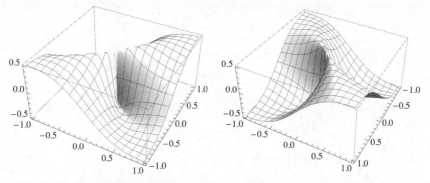

The surface indicates that the function is doing something very interesting around $(0, 0)$. Let's begin our investigation by looking at the behavior of $f(x, y)$ along lines $y = m x$:

```
f[x, m x] // Simplify
```

$$\frac{m}{1 + m^2}$$

So $f(x, y)$ is constant along any line $y = mx$ through the origin (except *at* the origin, where it is undefined), and the constant value is $m/(1 + m^2)$. (We need to think about the y-axis separately here: note that $f(0, y) = 0$, which turns out to be the limiting value of $m/(1 + m^2)$ as $m \to \infty$.) So we conclude that $\lim_{(x, y) \to (0, 0)} f(x, y)$ does not exist.

The fact that $f(x, y)$ is constant along lines through the origin means that these lines are contours of f. (Do you think that contours crossing at $(0, 0)$ has anything to do with the fact that $\lim_{(x, y) \to (0, 0)} f(x, y)$ does not exist?)

```
ContourPlot[f[x, y], {x, -1, 1}, {y, -1, 1},
    Contours → {-.3, 0, .3}, ContourShading → False]
```

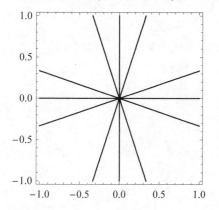

◆ Exercises

1. Make contour plots for each of the functions in Examples 4.1.1 and 4.1.2. Notice that contours intersect at the origin in each case. Explain why distinct contours can intersect only at a point where the limit of the function does not exist.

2. In a manner similar to Examples 4.1.1-4.1.3, investigate the behavior of the following functions at (0, 0). Include a contour plot.

a) $f(x, y) = \dfrac{x^2 - y^2}{x^4 + y^4}$ b) $f(x, y) = \dfrac{x^2 - y^2}{x^2 + y^2}$ c) $f(x, y) = \dfrac{\sin(x^2 + y^2)}{x^2 + y^2}$

d) $f(x, y) = \dfrac{\sin\sqrt{x^2 + y^2}}{x^2 + y^2}$ e) $f(x, y) = \sin\left(\dfrac{1}{x^2 + y^2}\right)$

4.2 Partial Derivatives

> Formal definition and detailed discussion of partial derivatives of multivariate functions are found in Section 15.3 of Stewart's *Calculus*. (ET:14.3)

The big question now is:

> *How might we define the derivative of a function of more than one variable?*

This is a complicated question, because it is not clear at this point what kind of beast such a thing should be. For a real-valued function f of one variable, you're probably used to thinking of f' as another function, whose values give the slope of the graph of f. But the graph of a function f of two variables is a surface in three-dimensional space, so what would we mean by slope?

Here is where we begin: Consider a function $f(x, y)$ of two variables, and note that if either of the two variables is held constant, then f becomes a function of just one variable— something we already know how to differentiate. This type of differentiation leads to two **partial derivatives**, one with respect to each variable.

■ **Example 4.2.1**

Consider the function

```
f[x_, y_] := 4 Exp[-(x² + y²) / 4] + y^2 / 6
```

Let's first look at the graph of f over $-3 \le x \le 3, \ -3 \le y \le 3$.

```
Plot3D[f[x, y], {x, -3, 3}, {y, -3, 3},
  BoxRatios → Automatic, ViewPoint → {6, 3, 2}]
```

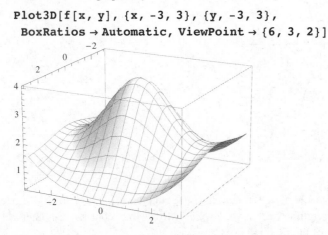

The grid curves in the surface plot are traces in $y = k$ and traces in $x = k$. What are the slopes of these traces? The slope of a trace in $y = k$ is the derivative of $f(x, k)$ with respect to x—or equivalently, the derivative of $f(x, y)$ with respect to x (with y treated as a constant) evaluated at $y = k$. The derivative of $f(x, y)$ with respect to x (with y treated as a constant) is called the **partial derivative of** $f(x, y)$ **with respect to** x, usually denoted by $\partial f / \partial x$ or $f_x(x, y)$. Here we have

```
xPartial[x_, y_] = ∂ₓ f[x, y]
```

$$-2 \, e^{\frac{1}{4}\left(-x^2-y^2\right)} \, x$$

For the sake of illustration, let's plot the trace in $y = 0.3$ along with its tangent line at $x = 0.5$.

```
tanLine[x_] = f[.5, .3] + xPartial[.5, .3] (x - .5)
Plot[{f[x, .3], tanLine[x]}, {x, -3, 3},
  PlotRange → {0, 5}, AxesLabel → {x, None}]
```

$$3.68905 - 0.918512 \, (-0.5 + x)$$

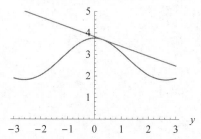

Similarly, the slope of a trace in $x = k$ is the derivative of $f(k, y)$ with respect to y—or equivalently, the derivative of $f(x, y)$ with respect to y (with x treated as a constant) evaluated at $x = k$. The derivative of $f(x, y)$ with respect to y (with x treated as a constant) is called the **partial derivative of** $f(x, y)$ **with respect to** y, usually denoted by $\partial f / \partial y$ or $f_y(x, y)$.

```
yPartial[x_, y_] = ∂ᵧ f[x, y]
```

$$\frac{y}{3} - 2 \, e^{\frac{1}{4}\left(-x^2-y^2\right)} \, y$$

Let's plot the trace in $x = .5$ along with its tangent line at $y = .3$.

```
tanLine[y_] = f[.5, .3] + yPartial[.5, .3] (y - .3)
Plot[{f[.5, y], tanLine[y]}, {y, -3, 3},
  PlotRange → {0, 5}, AxesLabel → {y, None}]
```

$$3.68905 - 0.451107 \, (-0.3 + y)$$

Now let's put all of this together in three dimensions. We'll first parametrize the traces in $y = 0.5$ and $x = 0.3$ as follows and plot them with `ParametricPlot3D`.

```
trace1 = ParametricPlot3D[{x, .5, f[x, .5]}, {x, -3, 3}];
trace2 = ParametricPlot3D[{.3, y, f[.3, y]}, {y, -3, 3}];
```

Then we parametrize the tangent lines and plot them along with the traces.

```
r0 = {.3, .5, f[.3, .5]}; v1 = yktrace'[.3]; v2 = xktrace'[.5];
Show[trace1, trace2,
  tan1 = ParametricPlot3D[r0 + x v1, {x, -3, 3}],
  tan2 = ParametricPlot3D[r0 + y v2, {y, -3, 3}],
  ViewPoint → {6, 3, 2}, PlotRange → {1, 5}]
```

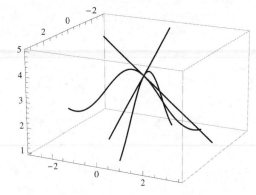

Finally, we recreate the surface and include that in the plot.

```
Show[trace1, trace2, tan1, tan2,
  Plot3D[f[x, y] - .015, {x, -3, 3}, {y, -3, 3},
    BoxRatios → Automatic, Mesh → None, PlotPoints → 20],
  ViewPoint → {6, 3, 2}, PlotRange → {0, 5} ]
```

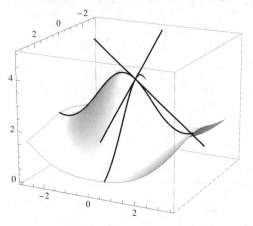

■ Higher-order Partial Derivatives

Just as in the case of a function of one variable, a function of two or more variables has second- (and higher-) order (partial) derivatives. A function of two variables has *four* second-order partial derivatives:

$$f_{xx}(x, y) = \frac{\partial}{\partial x} \frac{\partial f}{\partial x} \quad \text{(i.e., with respect to } x \text{ twice)}$$

$$f_{xy}(x, y) = \frac{\partial}{\partial y} \frac{\partial f}{\partial x} \quad \text{(i.e., first with respect to } x, \text{ then } y)$$

$$f_{yx}(x, y) = \frac{\partial}{\partial x} \frac{\partial f}{\partial y} \quad \text{(i.e., first with respect to } y, \text{ then } x)$$

$$f_{yy}(x, y) = \frac{\partial}{\partial y} \frac{\partial f}{\partial y} \quad \text{(i.e., with respect to } y \text{ twice)}$$

■ Example 4.2.2

Compute the two first-order and the four second-order partial derivatives of

$$f(x, y) = x \sin y + y \cos x.$$

These are the computations:

```
f[x_, y_] := x Sin[y] + y Cos[x]

{∂ₓ f[x, y], ∂_y f[x, y]}

{-y Sin[x] + Sin[y], Cos[x] + x Cos[y]}

{∂_{x,x} f[x, y], ∂_{x,y} f[x, y], ∂_{y,x} f[x, y], ∂_{y,y} f[x, y]}

{-y Cos[x], Cos[y] - Sin[x], Cos[y] - Sin[x], -x Sin[y]}
```

It may be slightly surprising that in the preceeding example the two "mixed partials" f_{xy} and f_{yx} turned out to be the same. It turns out, however, that this always happens—provided that the mixed partials are each continuous. This is *Clairaut's theorem*.

Let us also take notice of how *Mathematica* displays as output the partial derivatives of a function that has no definition. So let's

```
Clear[f]
```

and look at something like

```
∂_{x,x,y} f[x, y]

f^(2,1)[x, y]
```

Notice that the superscript in the output is an ordered pair (i, j) where i is the number of differentiations with respect to the first variable and j is the number of differentiations with respect to the second variable. In this expression:

$\partial_{s,s,t,u,s,u}$ `f[s, t, u]`

`f`$^{(3,1,2)}$ `[s, t, u]`

the superscript (3, 1, 2) indicates three differentiations with respect to the first variable, one differentiation with respect to the second variable, and two differentiations with respect to the third variable.

◆ Exercises

3. Create a plot like the final one in Example 4.2.1 for $f(x, y) = x^2 + y^2$, where the specified point is (.5, .5) (rather than (.5, .3)). Also, plot over $-1 \leq x \leq 1$ and $-1 \leq y \leq 1$ (rather than $-3 \leq x \leq 3$ and $-3 \leq y \leq 3$). Adjust other parameters to fine-tune the plot.

4. Verify that each of the following functions satisfies *Laplace's equation*:

$$u_{xx} + u_{yy} = 0.$$

(In other words, compute $u_{xx} + u_{yy}$ and observe that the result is zero.)

a) $u(x, y) = e^{-x} \sin y$ b) $u(x, y) = \tan^{-1}(y/x)$ c) $u(x, y) = \sqrt{x^2 + y^2}$

5. Verify that $u(t, x, y) = e^{-2t} \sin x \cos y$ satisfies the *heat equation*:

$$u_t - \left(u_{xx} + u_{yy}\right) = 0.$$

(In other words, compute $u_t - \left(u_{xx} + u_{yy}\right)$ and observe that the result is zero.)

6. Verify that, for *any* differentiable function f of one variable, the function of two variables given by $u(t, x) = f(x - t)$ satisfies the (one-dimensional) *wave equation*:

$$u_t + u_x = 0.$$

(In other words, compute $u_t + u_x$ and observe that the result is zero.) Then plot the surface $u(t, x) = f(x - t)$, where

`f[x_] := If[0 < x < `π`, Sin[x]`^{2}`, 0]`

over $0 \leq x \leq 10$ and $0 \leq t \leq 10$. (This is a plot of a "travelling wave." Keep in mind that the variable t represents time.)

4.3 The Tangent Plane and Linear Approximation

Tangent planes and linear approximation are the subject of Section 15.4 of Stewart's *Calculus*. (ET:14.4)

■ The Tangent Plane

At a given point on a smooth surface, there are infinitely many tangent lines. All of these tangent lines lie on a plane that we call the **tangent plane**.

Tangent vectors at $(a, b, f(a, b))$ are

```
Clear[f, a, b];
tanX[a_, b_] = {1, 0, ∂ₓ f[x, y]} /. {x → a, y → b}
tanY[a_, b_] = {0, 1, ∂ᵧ f[x, y]} /. {x → a, y → b}
```

$\{1, 0, f^{(1,0)}[a, b]\}$

$\{0, 1, f^{(0,1)}[a, b]\}$

A normal vector is given by their cross product:

```
nrml[a_, b_] = tanX[a, b] × tanY[a, b]
```

$\{-f^{(1,0)}[a, b], -f^{(0,1)}[a, b], 1\}$

The equation of the tangent plane at $(a, b, f(a, b))$ is then:

```
({x, y, z} - {a, b, f[a, b]}).nrml[a, b] == 0
```

$z - f[a, b] - (-b + y) f^{(0,1)}[a, b] - (-a + x) f^{(1,0)}[a, b] == 0$

Solving for z we find

```
z == (z /. First[Solve[%, z]])
```

$z == f[a, b] - b f^{(0,1)}[a, b] + y f^{(0,1)}[a, b] - a f^{(1,0)}[a, b] + x f^{(1,0)}[a, b]$

that is,

$$z = f(a, b) + f_x(a, b)(x - a) + f_y(a, b)(y - b).$$

(*Notice how this can be viewed as a sensible extension of the familiar equation*

$$y = f(a) + f'(a)(x - a)$$

of the tangent line to the graph of a function f of one variable at $(a, f(a))$.)

■ Example 4.3.1

Plot the surface $z = 3 - (x^2 + y^2)$ *along with the tangent plane at* $\left(\frac{1}{2}, \frac{1}{3}, f\left(\frac{1}{2}, \frac{1}{3}\right)\right)$. *Also show the tangent vectors* $\left\langle 1, 0, f_x\left(\frac{1}{2}, \frac{1}{3}\right)\right\rangle$ *and* $\left\langle 0, 1, f_y\left(\frac{1}{2}, \frac{1}{3}\right)\right\rangle$ *as well as the normal vector* **n** *resulting from their cross product.*

Here's a plot of the surface:

```
f[x_, y_] := 3 - (x² + y²);
surf = Plot3D[f[x, y], {x, -2, 2}, {y, -2, 2},
  BoxRatios → {1, 1, 1} , ViewPoint → {6, 2, 2}]
```

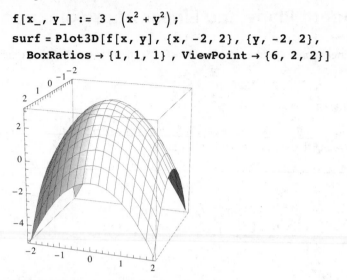

Using the functions we defined above, the tangent vectors we want are

```
a = 1 / 3;  b = 1 / 4;
tanX[a, b]
tanY[a, b]
```

$$\left\{1, 0, -\frac{2}{3}\right\}$$

$$\left\{0, 1, -\frac{1}{2}\right\}$$

and the resulting normal vector is

```
nrml[a, b]
```

$$\left\{\frac{2}{3}, \frac{1}{2}, 1\right\}$$

Having computed these vectors, the most convenient way to find the equation of the tangent plane is:

```
({x, y, z} - {a, b, f[a, b]}).nrml[a, b] == 0 // Simplify
```

$$24 (4 x + 3 y + 6 z) == 457$$

```
z[x_, y_] = z /. Solve[%, z][[1]]
```

$$\frac{1}{144} (457 - 96 x - 72 y)$$

Now the pictures. This creates a plot of the tangent and normal vectors:

```
pt = {a, b, f[a, b]};
vecs = Show[Vector /@
    {{pt, tanX[a, b]}, {pt, tanY[a, b]}, {pt, nrml[a, b]}}];
```

This creates a plot of the tangent plane:

```
tanPlane = Plot3D[z[x, y], {x, -1, 1.5},
   {y, -1, 1.5}, PlotStyle → Opacity[.5]];
```

This shows the vectors and the tangent plane along with the surface:

```
Show[surf, vecs, tanPlane,
 ViewPoint → {6, 3, 2}, PlotRange → {-1, 4.5} ]
```

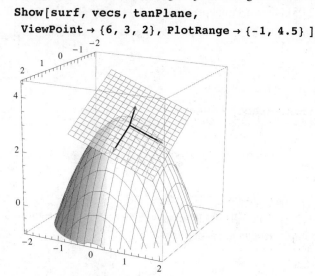

■ Linear Approximation

A good way to think of the tangent plane to the surface $z = f(x, y)$ at $(a, b, f(a, b))$ is that it is the graph of the linear function that best approximates $f(x, y)$ near (a, b). This linear function, denoted here by $\Lambda_{(a, b)}(x, y)$, is the **linear approximation**—or **linearization**—of f at (a, b). It is given by

$$\Lambda_{(a, b)}(x, y) = f(a, b) + f_x(a, b)(x - a) + f_y(a, b)(y - b).$$

A *Mathematica* definition of this is

```
Λ[{x_, y_}, {a_, b_}] := f[a, b] +
 (∂ₓf[x, y] /. {x → a, y → b}) (x - a) + (∂y f[x, y] /. {x → a, y → b}) (y - b)
```

(To get Λ (upper-case lambda), type ESC L ESC.)

■ Example 4.3.2

Find the linearization at (3, 2) *of*

```
f[x_, y_] := x^y - y^x
```

Use it to approximate $f(2.97, 2.05)$. *Also compute the relative error.*

The linearization at (3, 2) is

```
λ[x_, y_] = Λ[{x, y}, {3, 2}]
1 + (-3 + x) (6 - 8 Log[2]) + (-2 + y) (-12 + 9 Log[3])
```

The approximation to $f(2.97, 2.05)$ given by the linearization is

approx = λ[2.97, 2.05]

0.880731

The exact value of $f(2.97, 2.05)$ to six decimal places is

exact = f[2.97, 2.05]

0.882731

Finally, the relative error in the approximation is

relErr = Abs[(approx - exact) / exact]

0.00226548

◆ Exercises

7. Following Example 4.3.1, find and plot the tangent plane to the surface $z = xy$ at $(-1, -1)$ together with (i) the surface $z = xy$ and (ii) the pair of tangent vectors and resulting normal vector from which the tangent plane equation is derived.

8. Find the linearization of

$$\mathbf{f[x\_, y\_] := x^{-y^{-x^y}}}$$

at $(2, 1)$ and use it to approximate $f(1.94, 1.04)$. Compute the relative error in the approximation. *Note:* In order to understand how *Mathematica* interprets this function, look at the **InputForm** of the expression by entering **InputForm[f[x,y]]**.

9. The linearization of a function $f(x, y, z)$ at a point $P(a, b, c)$ is

$$\Lambda_P(x, y, z) = f(a, b, c) + f_x(a, b, c)(x - a) + f_y(a, b, c)(y - b) + f_z(a, b, c)(z - c).$$

Find the linearization of $f(x, y, z) = \sin(x \cos(y \sin z))$ at $\left(\frac{\pi}{3}, \frac{2\pi}{3}, \frac{\pi}{6}\right)$. Use it to approximate $f(1, 2, .5)$. Compute the relative error in the approximation.

4.4 Directional Derivatives and the Gradient

Section 15.6 of Stewart's *Calculus* contains formal definitions and thorough discussion of the directional derivative and the gradient vector. (ET:14.6)

■ Directional Derivatives

The (first-order) partial derivatives of a function give the rate of change in the function in each of the coordinate directions. For a function of two variables, say $f(x, y)$, this means that $f_x(a, b)$ is the rate of change in $f(x, y)$ at (a, b) in the direction of the vector $\mathbf{i} = \langle 1, 0 \rangle$, and $f_y(a, b)$ is the rate of change in $f(x, y)$ at (a, b) in the direction of the vector $\mathbf{j} = \langle 0, 1 \rangle$. (Keep in mind that these points and vectors live in the xy-plane, not on the surface.) The picture below shows these direction vectors in the xy-plane directly beneath a pair of tangent vectors whose slopes are given by $f_x(a, b)$ and $f_y(a, b)$. (Note that the code requires the Vectors package.)

```
f[x_, y_] := .5 + x^2 + y^2; a = 1 / 3; b = 1 / 4; pt = {a, b, f[a, b]};
tanX[x_, y_] = {1, 0, ∂_x f[x, y]}; tanY[x_, y_] = {0, 1, ∂_y f[x, y]};
Needs["Vectors`"]; Show[
 Plot3D[f[x, y], {x, -1, 2}, {y, -1, 2},
  BoundaryStyle → None, PlotStyle → Opacity[.7]],
 Vector /@ {{{a, b, 0}, {1, 0, 0}}, {{a, b, 0}, {0, 1, 0}},
   {pt, tanX[a, b]}, {pt, tanY[a, b]}}, ijk, ViewPoint → {6, 2, 2},
 BoxRatios → {3, 3, 2}, PlotRange → {-.2, 2}, Ticks → None]
```

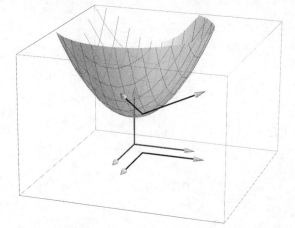

Given a unit vector $\mathbf{u} = \langle u_1, u_2 \rangle$, the **directional derivative** of f at (x, y) in the direction of $\mathbf{u}$ is

$$D_{\mathbf{u}} f(x, y) = f_x(x, y) u_1 + f_y(x, y) u_2.$$

(Note: *It is crucial that* $\mathbf{u}$ *be a unit vector.*) Computation is more convenient if we notice that $D_{\mathbf{u}} f(x, y)$ can be written as a dot product:

$$D_{\mathbf{u}} f(x, y) = \langle f_x(x, y), f_y(x, y) \rangle \cdot \mathbf{u}.$$

The directional derivative of a function of three variables in the direction of a unit vector $\mathbf{u} = \langle u_1, u_2, u_3 \rangle$ is given by an extension of the previous definition:

$$D_{\mathbf{u}} f(x, y, z) = f_x(x, y, z) u_1 + f_y(x, y, z) u_2 + f_z(x, y, z) u_3$$

$$= \langle f_x(x, y, z), f_y(x, y, z), f_z(x, y, z) \rangle \cdot \mathbf{u}.$$

■ Example 4.4.1

For the function

```
f[x_, y_] := x Cos[x y] + 3 (1 - Sin[x + y])
```

find the directional derivative at $(\pi/3, \pi/4)$ *in the direction of the vector* $\langle 2, 1 \rangle$.

The first thing we need to do is compute the *unit* direction vector:

```
v = {2, 1}; u = v / √v.v
```

$$\left\{ \frac{2}{\sqrt{5}}, \frac{1}{\sqrt{5}} \right\}$$

The partial derivatives of f are

```
{fx[x_, y_] = ∂x f[x, y], fy[x_, y_] = ∂y f[x, y]}
```

$$\left\{ \cos[x\,y] - 3\cos[x + y] - x\,y\,\sin[x\,y], \ -3\cos[x + y] - x^2\,\sin[x\,y] \right\}$$

Evaluated at $(\pi/3, \pi/4)$ these become

```
{fx[π / 3, π / 4], fy[π / 3, π / 4]}
% // N
```

$$\left\{ \frac{3\left(-1+\sqrt{3}\right)}{2\sqrt{2}} + \cos\left[\frac{\pi^2}{12}\right] - \frac{1}{12}\pi^2 \sin\left[\frac{\pi^2}{12}\right], \ \frac{3\left(-1+\sqrt{3}\right)}{2\sqrt{2}} - \frac{1}{9}\pi^2 \sin\left[\frac{\pi^2}{12}\right] \right\}$$

```
{0.854147, -0.0271772}
```

The desired directional derivative is

```
{fx[π / 3, π / 4], fy[π / 3, π / 4]}.u // Simplify
duf = % // N
```

$$\frac{9\left(9\sqrt{2}\left(-1+\sqrt{3}\right) + 8\cos\left[\frac{\pi^2}{12}\right]\right) - 10\,\pi^2 \sin\left[\frac{\pi^2}{12}\right]}{36\sqrt{5}}$$

```
0.751818
```

Now to illustrate what we've computed, let's plot the surface and include in the plot the following vectors:

$\langle u_1, u_2, 0 \rangle$ (that's $\mathbf{u}$ in the xy-plane) with initial point at $(\pi/3, \pi/4, 0)$,

$\langle u_1, u_2, D_{\mathbf{u}} f(\pi/3, \pi/4) \rangle$ with initial point at $(\pi/3, \pi/4, f(\pi/3, \pi/4))$.

The second of these vectors will be tangent to the surface and would coincide with **u** if it were projected onto the *xy*-plane.

```
Show[Plot3D[f[x, y], {x, 0, π/2}, {y, 0, π/2}],
 Vector /@ {{{π/3, π/4, 0}, Append[u, 0]},
  {{π/3, π/4, f[π/3, π/4]}, Append[u, duf]}}, ijk,
 PlotRange → {-.1, 2}, ViewPoint → {4, 1, 1}, BoxRatios → Automatic]
```

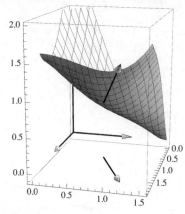

■ Example 4.4.2

Find the directional derivative of

$$f[x\_, y\_, z\_] := 1 \Big/ \sqrt{x^2 + y^2 + z^2}$$

at (3, 2, 1) *in the direction of the vector* ⟨1, −1, −1⟩.

As in the last example, the direction vector must be scaled to obtain a unit direction vector.

```
v = {1, -1, -1}; u = v / √(v.v)
```

$$\left\{ \frac{1}{\sqrt{3}}, -\frac{1}{\sqrt{3}}, -\frac{1}{\sqrt{3}} \right\}$$

The partial derivatives of f are

```
{fx[x_, y_, z_] = ∂x f[x, y, z], fy[x_, y_, z_] = ∂y f[x, y, z],
 fz[x_, y_, z_] = ∂z f[x, y, z]}
```

$$\left\{ -\frac{x}{\left(x^2+y^2+z^2\right)^{3/2}}, -\frac{y}{\left(x^2+y^2+z^2\right)^{3/2}}, -\frac{z}{\left(x^2+y^2+z^2\right)^{3/2}} \right\}$$

Evaluating these at (3, 2, 1) we obtain

```
{fx[3, 2, 1], fy[3, 2, 1], fz[3, 2, 1]}
```

$$\left\{ -\frac{3}{14\sqrt{14}}, -\frac{1}{7\sqrt{14}}, -\frac{1}{14\sqrt{14}} \right\}$$

So the desired directional derivative is

```
%.u
```

0

■ The Gradient

Let f be a function of two variables. The **gradient** of f at (x, y) is the vector

$$\nabla f(x, y) = \langle f_x(x, y), f_y(x, y) \rangle.$$

Similarly, the gradient at (x, y, z) of a function f of three variables is the vector

$$\nabla f(x, y, z) = \langle f_x(x, y, z), f_y(x, y, z), f_z(x, y, z) \rangle.$$

Note that with the gradient defined as such, we can express the definition of directional derivative (for a function of any number of variables) simply as

$$D_u f = \nabla f \cdot \mathbf{u}.$$

Moreover, a very important fact about the gradient is the following.

☐ *If $\nabla f \neq \mathbf{0}$, then ∇f gives the direction of greatest increase in f, and $-\nabla f$ gives the direction of greatest decrease in f.*

■ Example 4.4.3

Let $f(x, y) = x^2 + \sin y - xy$. Create a plot containing each of: (i) the surface $z = f(x, y)$ for $0 \leq x \leq 2$ and $0 \leq y \leq 2$; (ii) the gradient vector $\nabla f(1, 1)$ in the xy-plane with initial point $(1, 1)$; (iii) the vector tangent to the surface at $(1, 1, f(1, 1))$ whose vertical projection onto the xy-plane is $\nabla f(1, 1)$.

Let's first define the function and compute its gradient at (x, y).

```
f[x_, y_] := x² + Sin[y] - x y;
gradf[x_, y_] = {∂x f[x, y], ∂y f[x, y]}
```

```
{2 x - y, -x + Cos[y]}
```

This creates the graphic for the surface:

```
surf = Plot3D[f[x, y], {x, 0, 2}, {y, 0, 2},
    BoxRatios → Automatic, ViewPoint → {4, 3, 2}, Mesh → None];
```

The gradient vector $\nabla f(1, 1)$ in the xy-plane (in three dimensions) is

```
gradvec = Append[gradf[1, 1], 0] // N
```

```
{1., -0.459698, 0.}
```

The tangent vector that we need is obtained by appending the length of the gradient to the gradient vector:

```
tanvec = Append[gradf[1, 1], √(gradf[1, 1].gradf[1, 1])] // N
```

```
{1., -0.459698, 1.1006}
```

This creates a graphic containing the vectors:

```
vecs = Vector /@ {{{1, 1, 0}, gradvec}, {{1, 1, f[1, 1]}, tanvec}};
```

Now, finally, we show everything at once:

```
Show[surf, vecs, ijk,
  ViewPoint → {2, -2.5, 1}, BoxRatios → Automatic,
  PlotRange → {-.02, 2}, AxesLabel → {x, y, z}]
```

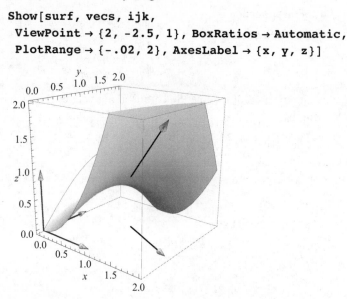

⚠ *Warning*: Don't confuse the gradient vector with the tangent vector to the surface. $\nabla f(x, y)$ lives in the *xy*-plane.

■ Example 4.4.4

Let f be the quadratic function

$$f[x\_, y\_, z\_] := (x - 2)^2 + (x - y - 1)^2 + (z - y)^2$$

and note that the minimum value of f occurs at (2, 1, 1).

a) *Find the rate of change in* $f(x, y, z)$ *in its direction of greatest decrease at the point* (3, 2, 0). *Then find a unit vector that points in that direction.*

b) *Plot the graph of f restricted to the line through* (3, 2, 0) *in the direction of greatest decrease there. Find the point on this line that minimizes f.*

(The repetition of this process is the *steepest descent method* for locating minima. See Project 7.11)

For part (a) we first need to compute the gradient at (x, y, z):

```
gradf[x_, y_, z_] = {∂_x f[x, y, z], ∂_y f[x, y, z], ∂_z f[x, y, z]}
```

$$\{2 (-2 + x) + 2 (-1 + x - y), -2 (-1 + x - y) - 2 (-y + z), 2 (-y + z)\}$$

The length of the gradient at (1, 2, 2) is

```
gradLen = √(gradf[3, 2, 0].gradf[3, 2, 0])
```

6

Therefore the rate of change in $f(x, y, z)$ in the direction of greatest decrease at $(3, 2, 0)$ is 6. The unit vector that points in that direction is

```
u = -gradf[3, 2, 0] / gradLen
```

$$\left\{ -\frac{1}{3}, -\frac{2}{3}, \frac{2}{3} \right\}$$

For part (b) we need a parametrization of the line. Since **u** is a unit vector, this one gives us a parametrization by arc length:

```
{x[t_], y[t_], z[t_]} = {3, 2, 0} + t u
```

$$\left\{ 3 - \frac{t}{3}, 2 - \frac{2 t}{3}, \frac{2 t}{3} \right\}$$

Here's the plot we want:

```
f[x[t], y[t], z[t]] // Simplify
Plot[f[x[t], y[t], z[t]], {t, -1, 4}]
```

$$5 - 6 t + 2 t^2$$

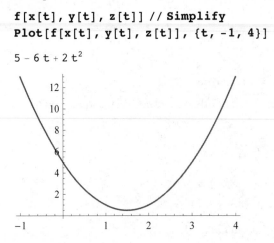

The minimum value of $f(x(t), y(t), z(t))$ occurs at $t = 3/2$; so the point on the line that minimizes f is

```
{x[t], y[t], z[t]} /. t → 3 / 2
```

$$\left\{ \frac{5}{2}, 1, 1 \right\}$$

■ Contours and the Gradient

Another very important property of the gradient vector is that it is always orthogonal to contours of the function.

 At any point P, $\nabla f(P)$ is orthogonal to the contour of f that passes through P.

Thus the gradient of any function f of two variables is orthogonal to the level curve $f(x, y) = f(x_0, y_0)$ at (x_0, y_0), and the gradient of a function of three variables is orthogonal to the level *surface* $f(x, y, z) = f(x_0, y_0, z_0)$ at (x_0, y_0, z_0).

Example 4.4.5

Consider the following function of two variables:

$$f[x\_, y\_] := x^4 - 3 \, x \, y + 2 \, y^2$$

The gradient is

$$gradf[x\_, y\_] = \left\{ \partial_x f[x, y], \partial_y f[x, y] \right\}$$

$$\left\{ 4 \, x^3 - 3 \, y, \, -3 \, x + 4 \, y \right\}$$

Contours of f are the level curves given by $f(x, y) = k$. To plot a contour that passes through a specific point (x_0, y_0), we simply need to plot the equation $f(x, y) = f(x_0, y_0)$. For instance, the following produces a plot of the contour through $(1/2, 1/2)$.

```
level = ContourPlot[f[x, y] == f[.5, .5],
    {x, 0, 1}, {y, 0, 1}, ContourShading → False]
```

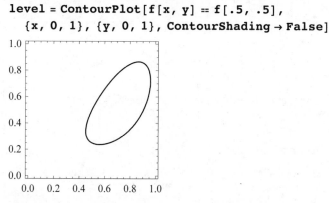

It's easy now to include the gradient vector at $(1/2, 1/2)$ in the plot:

```
Show[level, Vector[{{.5, .5}, gradf[.5, .5]}],
    PlotRange → All, AspectRatio → Automatic]
```

To plot several contours through a collection of specific points, we can create a table of function values to use with the Contours option. Let's say we want to use a few points along the line $y = .85 \, x$. The following does the job:

```
vals = Table[f[t, .85 t], {t, .5, 2, .25}];
levels = ContourPlot[f[x, y], {x, -1, 2.5}, {y, 0, 1.5},
    Contours → vals, AspectRatio → Automatic, ContourShading → False]
```

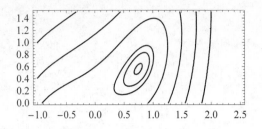

Now let's include a few gradient vectors in the plot:

```
Show[levels,
  Vector /@ Table[{{x, .85 x}, gradf[x, .85 x]}, {x, .5, 1, .25}]]
```

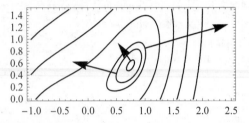

Note that each of the gradient vectors is orthogonal to the corresponding contour.

■ Example 4.4.6

Consider the following function of three variables:

```
f[x_, y_, z_] := x² + y² + z² + x y z
```

The gradient is

```
gradf[x_, y_, z_] = {∂ₓ f[x, y, z], ∂ᵧ f[x, y, z], ∂_z f[x, y, z]}
```

$$\{2\,x + y\,z,\ 2\,y + x\,z,\ x\,y + 2\,z\}$$

Let's plot the contour $f(x, y, z) = 2$.

```
r2 = √2 ; surf = ContourPlot3D[f[x, y, z] == 2,
  {x, -r2, r2}, {y, -r2, r2}, {z, -r2, r2}]
```

The following places several gradient vectors on the surface. As expected, these vectors are orthogonal to the surface.

```
Show[surf, Vector /@ {{{1, -1, 1}, gradf[1, -1, 1]},
    {{.5, -.93, 1.2}, gradf[.5, -.93, 1.2]},
    {{.25, -1.36, .5}, gradf[.25, -1.36, .5]},
    {{1.3, .25, -.68}, gradf[1.3, .25, -.68]}},
  ViewPoint → {1,-3,0}, Boxed → False, Axes → None, PlotRange → All]
```

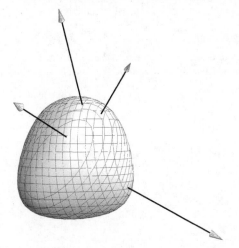

On Notation and Functions of n Variables

Regardless of the number of variables—whether it be two or three or more—the gradient of a function f is defined essentially the same way. If f is a function of n variables $x_1, x_2, \ldots, x_n$, then the gradient of f is

$$\nabla f (x_1, x_2, \ldots, x_n) = \left\langle \frac{\partial f}{\partial x_1}, \frac{\partial f}{\partial x_2}, \ldots, \frac{\partial f}{\partial x_n} \right\rangle.$$

When dealing in general terms with a function f of an unspecified number of variables, it is convenient to think of f as a function of just one variable χ that is viewed as either a vector with n components or a point in n dimensional space:

$$\chi = \langle x_1, x_2, \ldots, x_n \rangle \text{ or } (x_1, x_2, \ldots, x_n).$$

Thinking in these terms, a function value is written simply as $f(\chi)$, and the gradient of f at an arbitrary point becomes simply $\nabla f(\chi)$. Moreover, the directional derivative of f at χ in the direction of a unit vector **u** is

$$D_{\mathbf{u}} f(\chi) = \nabla f(\chi) \cdot \mathbf{u},$$

and the linearization of f at a point A (see Section 4.3) can be expressed as

$$\Lambda_P(\chi) = f(A) + \nabla f(A) \cdot (\chi - A).$$

This last formula expresses something quite complicated in a very concise, compact, easily-remembered form in which it looks very similar to the familiar analog for a function of one variable:

$$\Lambda_a(x) = f(a) + f'(a)\,(x - a)\,.$$

Can we take advantage of this with *Mathematica*? Yes, because *Mathematica* allows us to define functions with vectors (lists) as arguments.

■ Example 4.4.7

The function

```
f[x_?VectorQ] := 1 / Sqrt[x.x]
```

can be applied to a vector of any length:

```
x = {x, y}; f[x]
x = {x, y, z}; f[x]
x = {1, 2, 3, 4, 5}; f[x]
```

$$\frac{1}{\sqrt{x^2+y^2}}$$

$$\frac{1}{\sqrt{x^2+y^2+z^2}}$$

$$\frac{1}{\sqrt{55}}$$

Notice what happens if we try to evaluate f at something other than a vector:

```
f[{1, {2}}]
f[x + y]
```

```
f[{1, {2}}]

f[x + y]
```

Nothing happens, because of the ?VectorQ test in the definition of f.

■ Example 4.4.8

We can also define a single command that will compute the gradient of a function of any number of variables. This command will have two arguments. The first is the *expression* whose gradient we're after, and the second is a vector of variables.

```
grad[ξ_, x_?VectorQ] := Table[D[ξ, x[[i]]], {i, Length[x]}]
```

The gradient of the function defined in Example 4.4.6 (with three variables) is now computed by

```
x = {x, y, z}; grad[f[x], x]
```

$$\left\{-\frac{x}{\left(x^2+y^2+z^2\right)^{3/2}}, -\frac{y}{\left(x^2+y^2+z^2\right)^{3/2}}, -\frac{z}{\left(x^2+y^2+z^2\right)^{3/2}}\right\}$$

Here's another application of grad, this time with a function of two variables:

grad$\left[x^3\, y\, +\, Cos[y]\, ,\, \{x,\, y\}\right]$

$\{3\, x^2\, y,\, x^3\, -\, Sin[y]\}$

Notice that the variables in the expression and the variables in the variable list needn't coincide.

grad$\left[x^3\, y\, +\, Cos[y]\, ,\, \{x,\, t\}\right]$

$\{3\, x^2\, y,\, 0\}$

■ Example 4.4.9

Though it is somewhat more complicated to define than our grad command from Example 4.4.8, we can also create a single function that will produce the linearization of a function of any number of variables. Recall the formula:

$$\Lambda_P(\chi) = f(A) + \nabla f(A) \cdot (\chi - A) \,.$$

We will call the command linApprox. It will have three arguments:

(1) an expression ξ representing $f(\chi)$;

(2) the vector of variables χ;

(3) a vector a representing the point A at which we want to linearize $f(\chi)$.

Some important issues are:

- χ and a must be vectors *and* must have the same dimension;

- ξ is an *expression*—not a function—and so we have to compute its value using substitution rules;

- the gradient of ξ must be computed with respect to the variables in χ *before* substitution rules are applied to evaluate it at a.

With these issues in mind, we'll create the following command. Notice how each of the issues above is addressed.

```
linApprox[ξ_, χ_?VectorQ, a_?VectorQ] := If[Length[χ] == Length[a],
  Module[{subs = Table[χ[[i]] → a[[i]], {i, Length[χ]}]},
   (ξ /. subs) + (Evaluate[grad[ξ, χ]] /. subs).(χ - a)]]
```

The following are a few applications of linApprox.

linApprox$\left[x^2 / y^3,\, \{x,\, y\},\, \{2,\, 1\}\right]$

$4 + 4\, (-2 + x) - 12\, (-1 + y)$

$\chi = \{x,\, y,\, z\};$ linApprox$\left[1 \Big/ \sqrt{\chi \cdot \chi}\, ,\, \chi,\, \{2,\, 1,\, 2\}\right]$

$\dfrac{1}{3} - \dfrac{2}{27}\, (-2 + x) + \dfrac{1 - y}{27} - \dfrac{2}{27}\, (-2 + z)$

linApprox$[x\, y\, z,\, \{x,\, y,\, z\},\, \{2,\, 1,\, 3\}]$

$6 + 3\, (-2 + x) + 6\, (-1 + y) + 2\, (-3 + z)$

◆ Exercises

10. Find the directional derivative of each of the following functions at the specified point and in the direction of $\mathbf{u} = \langle \cos\theta, \sin\theta \rangle$. Plot the resulting function of θ for $0 \le \theta \le 2\pi$, and find where the extreme values occur. Then, for those values of θ, verify that $\mathbf{u}$ corresponds to either $\pm\nabla f$.

   a) $f(x, y) = \dfrac{1}{\sqrt{x^2+y^2}}$ at $(1, 1)$

   b) $f(x, y) = x^2 y^3$ at $(1, 1)$

   c) $f(x, y) = \cos xy$ at $(1/4, \pi)$

11. Find the directional derivative of each of the following functions at the specified point and in the direction of $\mathbf{u} = \langle \cos\theta\sin\phi, \sin\theta\sin\phi, \cos\phi \rangle$. Plot the resulting function of θ and ϕ for $0 \le \theta \le 2\pi$ and $0 \le \phi \le \pi$.

   a) $f(x, y, z) = \dfrac{1}{\sqrt{x^2+y^2+z^2}}$ at $(1, 1, 1)$

   b) $f(x, y, z) = x^2 y^3 z$ at $(1, 1, 1)$

   c) $f(x, y, z) = \cos xyz$ at $(1/2, \pi, 1/3)$

12. Carry out two more steps of the procedure in Example 4.4.4, each time starting with the point that resulted from the previous step. Also compute the value of f at each new point.

13. Give a simple geometric argument for why the gradient of $f(x, y) = x^2 + y^2$ at (a, b) is orthogonal to the contour of f through (a, b). Include an illustration consisting of a plot of the unit circle and several gradient vectors with initial points at the origin.

14. The `grad` function defined in this section can be applied to a vector containing several functions. The transpose of the result gives a list of gradients; for example:

    $$\texttt{grad}\big[\{\texttt{x}^2\,\texttt{y, Sin[x y], x / y}\}, \{\texttt{x, y}\}\big] \;\texttt{// Transpose}$$

    Since `grad` returns a vector, what happens if we apply `grad` to the result of `grad` and look at the transpose? For example, see what happens when you enter

    $$\texttt{grad}\big[\texttt{grad}\big[\texttt{x}^3\,\texttt{y}^2, \{\texttt{x, y}\}\big], \{\texttt{x, y}\}\big] \;\texttt{// Transpose // MatrixForm}$$

    What are the entries of the resulting matrix relative to the original function? This is the **Hessian matrix** of the function, typically denoted by either $\nabla^2 f(x, y)$ or $H(x, y)$. Note also that the result is the same even if we don't transpose the matrix. Why?

15. Let $f(x, y)$ be defined as follows.

    $$\texttt{f[x\_, y\_] := x y}^2 \texttt{/} \big(\texttt{x}^2 + \texttt{y}^4\big)$$

    a) Show that $f(x, y) \to 0$ as $(x, y) \to (0, 0)$ along *any* straight line, but $f(x, y) \to 1/2$ as $(x, y) \to (0, 0)$ along the parabola $x = y^2$. Conclude that f is not continuous at $(0, 0)$.

Suppose we redefine f as

$$f(x, y) = \begin{cases} \dfrac{x y^2}{x^2 + y^4} & \text{if } x^2 + y^2 \neq 0 \\ 0 & \text{if } x^2 + y^2 = 0 \end{cases}$$

Then f still is not continuous at $(0, 0)$, but it behaves as if it were continuous there when we approach $(0, 0)$ along straight lines. Use the definition of $f_x(0, 0)$ and $f_y(0, 0)$ to show that each exists and equals 0.

c) Note that if we were to use the formula involving the gradient to calculate directional derivatives at $(0, 0)$, we would find that $D_u f(0, 0) = 0$ in every direction. But because f is not continuous at $(0, 0)$, we cannot compute any directional derivative there with the formula that involves the gradient—we must use the basic definition:

$$D_u f(x, y) = \phi'(0), \text{ where } \phi(t) = f(x + t u_1, y + t u_2).$$

Find $D_u f(0, 0)$ for: (i) $\mathbf{u} = \langle 0, 1 \rangle$, and (ii) $\mathbf{u} = \langle \cos \theta, \sin \theta \rangle$, $\theta \neq \frac{\pi}{2}$. Conclude that the directional derivative of f exists at $(0, 0)$ *in every direction*, even though f is *not continuous* at $(0, 0)$!

4.5 Optimization and Lagrange Multipliers

Optimization is the subject of Section 15.7 in Stewart's Calculus. The method of Lagrange Multipliers is the subject of Section 15.8. (ET:14.7-8)

■ Critical Points

The problem that we are now concerned with is analogous to the familiar "max/min" problems from single-variable calculus. Given a function f of two or more variables, we wish to find the points in the domain of f at which f attains a local maximum or local minimum value.

The fundamental tool that we have to work with is this:

☐ *If f attains either a locally maximum or minimum value at a point χ at which f is differentiable (i.e., $\nabla f(\chi)$ exists), then $\nabla f(\chi) = 0$.*

Points χ at which either $\nabla f(\chi)$ doesn't exist or $\nabla f(\chi) = 0$ are the **critical points** of f.

■ Example 4.5.1

Find all the critical points of

```
f[x_, y_] := x^4 - 3 x y + 2 y^2
```

Use the graph of f to help determine whether the value of f at each critical point is a local maximum, a local minimum, or neither.

Let's first find the critical points. (The uses the function `grad` that was defined in the preceding section.)

```
grad[f[x, y], {x, y}]
{x, y} /. Solve[% == {0, 0}, {x, y}]
```

$$\left\{4\,x^3 - 3\,y,\ -3\,x + 4\,y\right\}$$

$$\left\{\left\{-\frac{3}{4},\ -\frac{9}{16}\right\},\ \{0,\ 0\},\ \left\{\frac{3}{4},\ \frac{9}{16}\right\}\right\}$$

So we have three critical points. Let's now get a good plot of the graph over a region that includes all three critical points.

```
Plot3D[f[x, y], {x, -3/2, 3/2}, {y, -3/2, 3/2},
  BoxRatios → {3, 3, 2}, ClippingStyle → None,
  PlotRange → {-.5, 1}, ViewPoint → {4, -2, 2.5}]
```

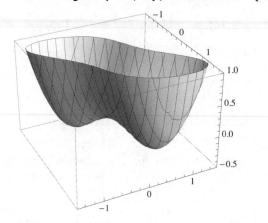

The graph makes it clear that f attains local minima at $\left(-\frac{3}{4}, -\frac{9}{16}\right)$ and $\left(\frac{3}{4}, \frac{9}{16}\right)$ and that the value at $(0, 0)$ is neither a local maximum nor a local minimum. (A critical point such as $(0, 0)$ in this example is called a *saddle point*.) Note that f has the same value at each of $\left(-\frac{3}{4}, -\frac{9}{16}\right)$ and $\left(\frac{3}{4}, \frac{9}{16}\right)$:

```
f[-3/4, -9/16]
f[3/4, 9/16]
% // N
```

$$-\frac{81}{256}$$

$$-\frac{81}{256}$$

```
-0.316406
```

This turns out also to be the *absolute* minimum value of f.

■ The Second Derivative Test

This is the two-variable analog to the second derivative test for functions of one variable:

☐ Suppose that $\nabla f(a, b) = 0$, and let $D = f_{xx}(a, b) f_{yy}(a, b) - f_{xy}(a, b)^2$.

- If $D > 0$ and $f_{xx}(a, b) > 0$, then f attains a local minimum at (a, b).

- If $D > 0$ and $f_{xx}(a, b) < 0$, then f attains a local maximum at (a, b).

- If $D < 0$, then f has neither a local maximum nor local minimum at (a, b).

- If $D = 0$, then f might attain a local minimum, a local maximum, or neither at (a, b). In other words, the test is inconclusive.

The quantity D here may be viewed as a *discriminant*. Here is a *Mathematica* definition:

```
disc[ξ_, {x_, y_}, {a_, b_}] := ∂x,x ξ ∂y,y ξ - (∂x,y ξ)² /. {x → a, y → b}
```

■ Example 4.5.2

Each of the following quadratic functions has a single critical point at $(0, 0)$. Plot each graph after deciding whether $f(0, 0)$ is a maximum, minimum, or neither.

```
f1[x_, y_] := x² - 2 x y + 2 y²;
f2[x_, y_] := 2 x² - 5 x y + y²;
f3[x_, y_] := -x² + x y - y²;
```

The first function f1 has a minimum at $(0, 0)$ because

```
disc[f1[x, y], {x, y}, {0, 0}]
```

```
4
```

is positive and $f_{xx}(0, 0) = 2 > 0$. The second function f2 has neither a maximum nor a minimum at $(0, 0)$ because

```
disc[f2[x, y], {x, y}, {0, 0}]
```

```
-17
```

is negative. The third function f3 has a maximum at $(0, 0)$ because

```
disc[f3[x, y], {x, y}, {0, 0}]
```

```
3
```

is positive and $f_{xx}(0, 0) = -1 < 0$. Here are the plots:

```
makePlot[fn_, opts___] := Plot3D[fn, {x, -1, 1}, {y, -1, 1},
    opts, ClippingStyle → None, Axes → None, BoxRatios → {1, 1, 1}]
```

```
GraphicsRow[{makePlot[f1[x, y], PlotRange → {0, 1}],
    makePlot[f2[x, y], PlotRange → {-1, 1}],
    makePlot[f3[x, y], PlotRange → {-1, 0}]}]
```

■ Example 4.5.3

Find all critical points of the function

$$f[x\_, y\_] := 3\,x - x^3 - 2\,x\,y^2 + 2\,y$$

Then at each critical point, determine whether f has a local maximum, local minimum, or neither. Finally, plot the surface to confirm the results.

First off, the gradient is

```
gradf = grad[f[x, y], {x, y}]
```

$$\{3 - 3\,x^2 - 2\,y^2,\ 2 - 4\,x\,y\}$$

The critical points can be found with Solve. Since the exact forms are rather messy, we'll also compute numerical values. For later convenience, we will also leave the result in the form of a list of rules.

```
critpts = Solve[gradf == {0, 0}, {x, y}]
% // N
```

$$\left\{\left\{x \to \frac{1}{6}\left(-3\sqrt{3-\sqrt{3}} - \sqrt{3\left(3-\sqrt{3}\right)}\right),\ y \to -\frac{1}{2}\sqrt{3-\sqrt{3}}\ \right\},\right.$$

$$\left\{x \to \frac{1}{3}\left(\frac{3\sqrt{3-\sqrt{3}}}{2} + \frac{1}{2}\sqrt{3\left(3-\sqrt{3}\right)}\right),\ y \to \frac{\sqrt{3-\sqrt{3}}}{2}\right\},$$

$$\left\{x \to \frac{1}{3}\left(-3\sqrt{\frac{3}{4}+\frac{\sqrt{3}}{4}} + \sqrt{3\left(\frac{3}{4}+\frac{\sqrt{3}}{4}\right)}\right),\ y \to -\sqrt{\frac{3}{4}+\frac{\sqrt{3}}{4}}\right\},$$

$$\left.\left\{x \to \frac{1}{6}\left(3\sqrt{3+\sqrt{3}} - \sqrt{3\left(3+\sqrt{3}\right)}\right),\ y \to \sqrt{\frac{3}{4}+\frac{\sqrt{3}}{4}}\right\}\right\}$$

$$\{\{x \to -0.888074,\ y \to -0.563016\},\ \{x \to 0.888074,\ y \to 0.563016\},$$
$$\{x \to -0.459701,\ y \to -1.08766\},\ \{x \to 0.459701,\ y \to 1.08766\}\}$$

We'll now compute a list containing the value of D at each critical point and a corresponding list containing the value of f_{xx} at each critical point. (The commands take advantage of the fact that `critpts` is a list of rules.)

```
Table[disc[f[x, y], {x, y}, {x, y}] /. critpts[[i]], {i, 1, 4}] // N
Table[∂x,x f[x, y] /. critpts[[i]], {i, 1, 4}] // N
```

```
{13.8564, 13.8564, -13.8564, -13.8564}
{5.32844, -5.32844, 2.75821, -2.75821}
```

From the numbers in the lists, we can make the following conclusions. Each of D and f_{xx} is positive at the first critical point,

```
{x, y} /. critpts[[1]] // N
```

```
{-0.888074, -0.563016}
```

So f has a local minimum there. At the second critical point,

```
{x, y} /. critpts[[2]] // N
```

```
{0.888074, 0.563016}
```

D is positive and f_{xx} is negative. So f has a local maximum there. At each of the third and fourth critical points,

```
{{x, y} /. critpts[[3]], {x, y} /. critpts[[4]]} // N
```

```
{{-0.459701, -1.08766}, {0.459701, 1.08766}}
```

D is negative. So each of those critical points is a saddle point. The following is a list of the function values at the four critical points.

```
Table[f[x, y] /. critpts[[i]], {i, 1, 4}] // N
```

```
{-2.52684, 2.52684, -2.36962, 2.36962}
```

Here is the surface near the four critical points:

```
Plot3D[f[x, y], {x, -3, 3}, {y, -3, 3},
  PlotRange → {-5, 5}, PlotPoints → 40, ViewPoint → {5, -1, 1},
  BoxRatios → {4, 4, 3}, ClippingStyle → None]
```

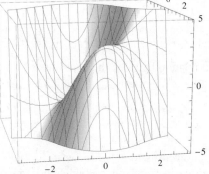

The local maximum at the second critical point is clearly in view, as are the two saddle points. The local minimum at the first critical point is hidden, but nevertheless evident from the shape of the surface.

■ Contours and Critical Points

Contours often provide information about the location of critical points and the behavior of the function near them. To illustrate this, let's just plot contours for each of the functions in Examples 4.5.2 and 4.5.3. Contour plots of the quadratic functions in Example 4.5.2 are as follows.

```
GraphicsRow[
  {ContourPlot[f1[x, y], {x,-1, 1}, {y,-1, 1}, ContourShading→False],
   ContourPlot[f2[x, y], {x,-1, 1}, {y,-1, 1}, ContourShading→False],
   ContourPlot[f3[x, y], {x,-1, 1}, {y,-1, 1}, ContourShading→False]}]
```

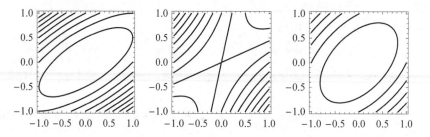

A contour plot of the function in Example 4.5.3 is shown below.

```
ContourPlot[f[x, y], {x, -1.5, 1.5}, {y, -1.5, 1.5},
  Contours → 81, ContourShading → False]
```

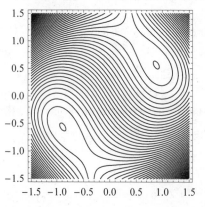

For each of these contour plots, study the correlation between the appearance of the contour plot near each critical point and whether the each critical point maximizes the function, minimizes the function, or is a saddle point.

■ FindRoot and FindMinimum

Up to this point, all of our examples have allowed us to solve for critical points with the Solve command. However, Solve is primarily designed for algebraic equations, and many (most?) problems involve functions whose gradients have non-algebraic components containing exponential, logarithmic, or trigonometric functions.

Mathematica has two built-in commands that can be brought to bear on such problems. FindRoot can be applied to the gradient to find critical points, and FindMinimum can be applied to the objective function to find the critical points at which local minima occur. FindRoot is based upon Newton's method for finding zeros of a function, and FindMinimum is based on the *steepest descent method*, to which we alluded in Example 4.4.4.

Each of FindRoot and FindMinimum requires us to supply it with an initial guess at the critical point we wish to locate. (This should sound familiar from what you know about Newton's method.) The following examples illustrate the use of each of these commands.

■ Example 4.5.4

Use FindRoot *to find all the critical points of*

$$f[x\_, y\_] := \left(y^2 - x + x\,y + Log\left[x^2 + y^4 + 1\right]\right) Exp\left[-\left(x^2 + y^2\right)\right]$$

in the square region where $-2 \le x \le 2$ *and* $0 \le y \le 2$.

Let's first plot the surface $z = f(x, y)$ and a contour plot.

```
surf =
  Plot3D[f[x, y], {x, -2, 2}, {y, -2, 2.}, BoxRatios → {1, 1, 1}];
cntrs = ContourPlot[f[x, y], {x, -2, 2}, {y, -2, 2.}, Contours → 20,
  ContourShading → False]; GraphicsRow[{surf, cntrs}]
```

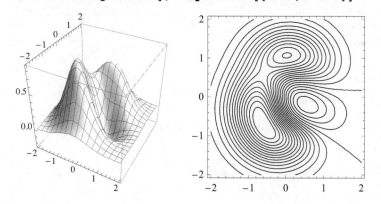

These plots indicate that f has:

   (i) a local maximum somewhere near $(-.5, -.75)$;

  (ii) a local maximum somewhere near $(0, 1)$;

 (iii) a local minimum somewhere near $(.5, -.2)$;

 (iv) a saddle point somewhere near $(-.5, .5)$.

Now let's find the gradient of f:

```
gradf = grad[f[x, y], {x, y}] // Simplify
```

$$\left\{ e^{-x^2-y^2} \left(-1 + y + \frac{2\,x}{1+x^2+y^4} - 2\,x\left(x\,(-1+y)+y^2+Log\left[1+x^2+y^4\right]\right)\right), \right.$$

$$\left. e^{-x^2-y^2} \left(x + 2\,y + \frac{4\,y^3}{1+x^2+y^4} - 2\,y\left(x\,(-1+y)+y^2+Log\left[1+x^2+y^4\right]\right)\right)\right\}$$

This locates the critical point near $(-.5, -.75)$ and computes the function value there:

```
FindRoot[gradf, {x, -.5}, {y, -.75}]
f[x, y] /. %
```

$\{x \rightarrow -0.597067, y \rightarrow -0.729097\}$

```
0.846801
```

This locates the critical point near $(0, 1)$ and computes the function value there:

```
FindRoot[gradf, {x, 0}, {y, 1}]
f[x, y] /. %
```

$\{x \rightarrow 0.023479, y \rightarrow 1.07379\}$

```
0.631213
```

This locates the critical point near $(.5, -.2)$ and computes the function value there:

```
FindRoot[gradf, {x, .5}, {y, -.2}]
f[x, y] /. %
```

$\{x \rightarrow 0.527469, y \rightarrow -0.188337\}$

```
-0.252022
```

This locates the critical point near $(-.5, .5)$ and computes the function value there:

```
FindRoot[gradf, {x, -.5}, {y, .5}]
f[x, y] /. %
```

$\{x \rightarrow -0.59617, y \rightarrow 0.606149\}$

```
0.485992
```

▪ Example 4.5.5

Use FindMinimum *to locate all the critical points of*

$$f[x\_, y\_] := \left(2y - x + Exp[1 - x^2 - y^2]\right) / \left(1 + x^2 + y^2\right)$$

in the square region where $-2 \leq x \leq 2$ *and* $0 \leq y \leq 2$.

Let's first plot the surface $z = f(x, y)$ and a contour plot.

```
GraphicsRow[
  {Plot3D[f[x, y], {x, -2, 2}, {y, -2, 2}, BoxRatios → {1, 1, 1},
    ViewPoint → {2, -4, 2}], ContourPlot[f[x, y], {x, -2, 2},
    {y, -2, 2}, Contours → 17, ContourShading → False]}]
```

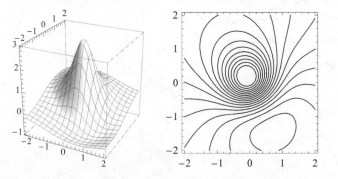

These plots indicate that f has:

(i) a local minimum somewhere near $(0.8, -1.5)$;

(ii) a local maximum somewhere near $(-0.1, 0.2)$.

This finds the critical point near $(0.8, -1.5)$ and the local minimum that f attains there:

```
FindMinimum[f[x, y], {x, .8}, {y, -1.5}]
```
```
{-0.946503, {x → 0.700885, y → -1.40177}}
```

To find a local *maximum* of f, we can apply `FindMaximum`. This finds the critical point near $(-.1, .2)$ and the local maximum that f attains there:

```
FindMaximum[f[x, y], {x, -.1}, {y, .2}]
```
```
{2.94127, {x → -0.0900656, y → 0.180131}}
```

■ Constrained Optimization and Lagrange Multipliers

Consider the following general problem, where f and g are two given differentiable functions of n variables and k is a given constant.

Find the extreme values of $f(\chi)$ over all χ satisfying $g(\chi) = k$.

This is a *constrained* optimization problem. The function f is the *objective function*, and $g(\chi) = k$ is the **constraint**.

If f and g are functions of two variables, this problem amounts to finding the extreme values of f on a level curve of g. If f and g are functions of three variables, the problem is to find the extreme values of f on a level surface of g.

⬚ At any local maximum or minimum value of $f(\chi)$ restricted to the contour $g(\chi) = k$, the gradient of f is orthogonal to that contour and therefore parallel to the gradient of g. Consequently, there is a number λ such that

$$\nabla f(\chi) = \lambda \nabla g(\chi) \text{ and } g(\chi) = k.$$

The number λ is called a *Lagrange multiplier*. Note that if f and g are functions of n variables, then these two equations actually represent a system of $n + 1$ equations in $n + 1$ unknowns: λ and the n variables in the vector χ.

■ Example 4.5.6

Find the minimum and maximum values of

```
f[x_, y_] := 2 x³ - 3 y + x
```

on the unit circle—that is, subject to the constraint $g(x, y) = 1$, where

```
g[x_, y_] := x² + y²
```

The system of equations that needs to be solved is

```
system = Flatten[
    {Thread[grad[f[x, y], {x, y}] == λ grad[g[x, y], {x, y}]],
    g[x, y] == 1}]
```

$\{1 + 6 x^2 == 2 x \lambda, -3 == 2 y \lambda, x^2 + y^2 == 1\}$

(`Thread` is used here to "thread" `Equal` over a list. See `Thread` in the documentation.)

Because these are fairly simple algebraic equations, `NSolve` will do the job nicely.

```
NSolve[system, {x, y, λ}]
```

$\{\{x \to -0.884839, y \to 0.465898, \lambda \to -3.21959\},$
$\{x \to 0.884839, y \to -0.465898, \lambda \to 3.21959\},$
$\{x \to -0.255169 + 0.351066 \, i, y \to 1.03231 + 0.0867771 \, i,$
$\lambda \to -1.44285 + 0.121288 \, i\}, \{x \to -0.255169 - 0.351066 \, i,$
$y \to 1.03231 - 0.0867771 \, i, \lambda \to -1.44285 - 0.121288 \, i\},$
$\{x \to 0.255169 + 0.351066 \, i, y \to -1.03231 + 0.0867771 \, i,$
$\lambda \to 1.44285 + 0.121288 \, i\}, \{x \to 0.255169 - 0.351066 \, i,$
$y \to -1.03231 - 0.0867771 \, i, \lambda \to 1.44285 - 0.121288 \, i\}\}$

Only the first two solutions are real. The values of f at the resulting points are:

```
f[x, y] /. %[[{1, 2}]]
```

$\{-3.66808, 3.66808\}$

Thus we see that (to six significant figures) the minimum value of f on the unit circle is -3.66808 and occurs at $(0.465898, -0.884839)$; the maximum is 3.66808 and occurs at $(-0.465898, 0.884839)$.

The following is plot of several contours f, including ones through the points found above, overlaid with the unit circle. Observe that the extreme values of f occur where the circle is tangent to the contour it intersects. (Hence the gradients are parallel.)

```
Show[ContourPlot[f[x, y], {x, -1.5, 1.5},
    {y, -1.5, 1.5}, Contours → {-6, -3.67, -2, 0, 2, 3.67, 6},
    ContourShading → False, ContourStyle → Thickness[.005]],
    ParametricPlot[{Cos[t], Sin[t]}, {t, 0, 2 π}]]
```

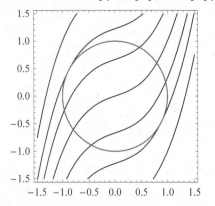

Example 4.5.7

A box with no top is to be constructed with a surface area 10 square feet. Find the dimensions that maximize the volume of the box.

Here our objective function is the volume of the box:

```
f[x_, y_, z_] := x y z
```

The constraint is $g(x, y, z) = 10$, where $g(x, y, z)$ is the surface area:

```
g[x_, y_, z_] := x y + 2 x z + 2 y z
```

The system of equations that needs to be solved is

```
system = {Thread[grad[f[x, y, z], {x, y, z}] ==
    λ grad[g[x, y, z], {x, y, z}]], g[x, y, z] == 10} // Flatten
```

$\{y\,z == (y + 2\,z)\,\lambda,\ x\,z == (x + 2\,z)\,\lambda,$
$\ x\,y == (2\,x + 2\,y)\,\lambda,\ x\,y + 2\,x\,z + 2\,y\,z == 10\}$

Solve will find the solutions of the system. (Here we will display both exact and numerical results.)

```
soln = Solve[system, {x, y, z, λ}]
% // N
```

$$\left\{\left\{x \to -\sqrt{\tfrac{10}{3}},\ y \to -\sqrt{\tfrac{10}{3}},\ z \to -\sqrt{\tfrac{5}{6}},\ \lambda \to -\tfrac{\sqrt{\tfrac{5}{6}}}{2}\right\},\right.$$

$$\left.\left\{x \to \sqrt{\tfrac{10}{3}},\ y \to \sqrt{\tfrac{10}{3}},\ z \to \sqrt{\tfrac{5}{6}},\ \lambda \to \tfrac{\sqrt{\tfrac{5}{6}}}{2}\right\}\right\}$$

$\{\{x \to -1.82574,\ y \to -1.82574,\ z \to -0.912871,\ \lambda \to -0.456435\},$
$\ \{x \to 1.82574,\ y \to 1.82574,\ z \to 0.912871,\ \lambda \to 0.456435\}\}$

The first of these solutions involves negative quantities that obviously are of no interest. So the maximum volume occurs at

```
{x, y, z} /. soln[[2]]
% // N
```

$$\left\{\sqrt{\tfrac{10}{3}},\ \sqrt{\tfrac{10}{3}},\ \sqrt{\tfrac{5}{6}}\right\}$$

$\{1.82574,\ 1.82574,\ 0.912871\}$

So the base of the optimal box is square, the height is half the edge-length of the base, and the volume of the box (in cubic feet) is

```
f[x, y, z] /. soln[[2]]
% // N
```

$$\frac{5\sqrt{\tfrac{10}{3}}}{3}$$

3.0429

Example 4.5.8

Find the minimum and maximum values of

$$f[x\_, y\_] := x\, y\, /\, 2 + Exp\left[-x^2 - y^2\right] Cos\left[\pi\, \left(x^2 - y\right)\right]$$

on the unit circle—that is, subject to the constraint g(x, y) = 1, where

$$g[x\_, y\_] := x^2 + y^2$$

Let's start with a plot of the surface $z = f(x, y)$:

```
Plot3D[f[x, y], {x, -1.5, 1.5}, {y, -1.5, 1.5},
   PlotRange → All, BoxRatios → Automatic, ViewPoint → {5, 1, 2}]
```

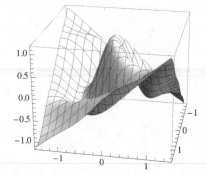

The system of equations we need to solve is

```
sys = {Thread[grad[f[x, y], {x, y}] == λ grad[g[x, y], {x, y}]],
   g[x, y] == 1} // Flatten
```

$$\left\{\frac{y}{2} - 2\, e^{-x^2-y^2}\, x\, Cos\left[\pi\,\left(x^2-y\right)\right] - 2\, e^{-x^2-y^2}\, \pi\, x\, Sin\left[\pi\,\left(x^2-y\right)\right] == 2\, x\, \lambda,\right.$$
$$\left.\frac{x}{2} - 2\, e^{-x^2-y^2}\, y\, Cos\left[\pi\,\left(x^2-y\right)\right] + e^{-x^2-y^2}\, \pi\, Sin\left[\pi\,\left(x^2-y\right)\right] == 2\, y\, \lambda,\ x^2 + y^2 == 1\right\}$$

In order to use `FindRoot` to locate the solutions, we need initial guesses for *x*, *y*, and *λ* at each solution. A contour plot of *f* overlaid with the unit circle will helpful for this.

```
Show[ContourPlot[f[x, y], {x, -1.1, 1.1},
   {y, -1.1, 1.1}, Contours → 18, ContourShading → False,
   ContourStyle → Thickness[.004]], ContourPlot[g[x, y] == 1,
   {x, -1.1, 1.1}, {y, -1.1, 1.1}, ContourShading → False]]
```

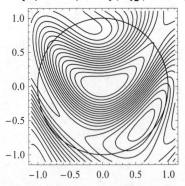

We can get rough estimates of the *x* and *y* values we're looking for by estimating where the contours of *f* are tangent to the circle. It is difficult to predict the corresponding values of λ; so we'll first use $\lambda = 0$ in each case and see what happens.

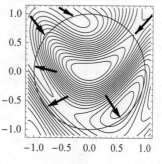

Close examination of the plot, starting in the lower-left quadrant of the plot and moving clockwise around the circle, leads to the following initial guesses:

```
guesses = {{-.75, -.5, 0}, {-.9, .1, 0}, {-.75, .6, 0},
    {-.25, .9, 0}, {.7, .7, 0}, {.6, -.75, 0}};
```

Let's see if those guesses work.

```
solns = Table[FindRoot[sys, {x, guesses[[i, 1]]},
    {y, guesses[[i, 2]]}, {λ, guesses[[i, 3]]}], {i, 1, 6}]
```

$\{\{x \to -0.795499, y \to -0.605955, \lambda \to 1.24747\},$
$\{x \to -0.994593, y \to 0.10385, \lambda \to -0.0891063\},$
$\{x \to -0.793211, y \to 0.608946, \lambda \to -0.632492\},$
$\{x \to -0.298239, y \to 0.954491, \lambda \to 0.00923684\},$
$\{x \to 0.780122, y \to 0.625628, \lambda \to -0.105032\},$
$\{x \to 0.561764, y \to -0.827297, \lambda \to 0.464783\}\}$

So our guesses seem to have worked well. Let's check by computing $\nabla f - \lambda \nabla g$ at each result.

```
grad[f[x, y], {x, y}] - λ grad[g[x, y], {x, y}] /. solns
```

$\{\{-6.66134 \times 10^{-16}, -6.66134 \times 10^{-16}\}, \{9.4369 \times 10^{-16}, 1.66533 \times 10^{-16}\},$
$\{2.22045 \times 10^{-16}, 5.55112 \times 10^{-17}\}, \{0., 1.66533 \times 10^{-16}\},$
$\{-5.55112 \times 10^{-16}, 4.44089 \times 10^{-16}\}, \{1.11022 \times 10^{-16}, 0.\}\}$

Here's a plot showing the points we've found:

```
Show[fig, Graphics[{PointSize[.03], Red, Point[{x, y}] /. solns}]]
```

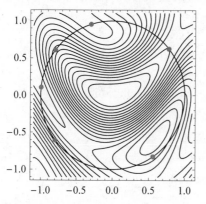

Finally, we'll compute the resulting values of f,

> ```
> f[x, y] /. solns
> ```
> {-0.0281226, -0.395924, 0.125625, -0.477878, 0.611385, -0.563811}

and conclude that the minimum is −0.563811 and the maximum is 0.611385.

◆ Exercises

For each of the functions in Exercises 16-18:

a) Find all critical points (using `Solve`).

b) Produce a surface plot (or plots) showing whether the function attains a local maximum, minimum, or neither at each critical point.

c) Use the second derivative test to confirm what is indicated by the plots from part (b).

d) Create a contour plot about each critical point (or all critical points) that substantiates the conclusions of parts (b) and (c). Explain.

16. $f(x, y) = x^4 - 8x^3 + 25x^2 - 32x - 6xy + 9y^2$

17. $f(x, y) = x^4 - 3x^2 - 6xy + 9y^2$

18. $f(x, y) = x - x^3 - 3xy^2 + y$

For each of the functions in Exercises 19-20:

a) Create a contour plot and a surface plot over the square described by $-2 \le x \le 2$ and $-2 \le y \le 2$.

b) Using initial guesses obtained from the contour plot, use `FindRoot` to find all the critical points at which f attains local minima and local maxima in that square.

c) Repeat part (b) using `FindMinimum`.

19. $f(x, y) = \left(x^2 - xy + y^2 - y\right) e^{-\left(x^2 + y^2\right)}$

20. $f(x, y) = \left(x^2 + xy + y^2 - x\right) \cos\left(\pi \sqrt{x^2 + y^2 + 1}\right)$

21. Using the Lagrange multipliers method, find the minimum and maximum values of each of the functions in 16-20 on:

a) the unit circle;   b) the ellipse $2x^2 + 3y^2 = 1$.

22. Find the minimum value of $f(x, y, z, w) = 4x^2 + 3y^2 + 2z^2 + w^2$, subject to the constraint $x + 2y + 3z + 4w = 100$,

a) using the Lagrange multipliers method;   b) without Lagrange multipliers.

23. Find the minimum value of $f(x, y) = x^2 + y^2$ on the curve $x^3 - 2x - 2y^3 + 3y = 3$. Create a plot showing the curve along with contours of f.

24. Find the minimum value of $f(x, y) = x^2 - x - y + y^2$ on the curve $x^3 - 2x - 2y^3 + 3y = 0$. Create a plot showing the curve along with contours of f.

5 Multiple Integrals

As you are probably already aware, one of the most powerful capabilities of *Mathematica* is its ability to compute integrals, both symbolically and numerically. In this chapter we will use *Mathematica* to explore *multiple integrals*—in particular, *double integrals* of functions of two variables and *triple integrals* of functions of three variables. Multiple integrals and their applications are the subject of Chapter 16 of Stewart's *Calculus*.

5.1 Double Integrals

See Sections 16.1-3 of Stewart's *Calculus* for detailed discussion of double integrals. (ET:15.1-3)

■ Midpoint Approximations

Just as in the case of a function of one variable, the (definite) integral of a function of two variables is defined as a limit of Riemann sums, and reasonably good approximations can be computed by means of a midpoint rule. For integrals of functions of two variables, this involves summing up function values computed at midpoints of rectangles, multiplied by rectangle area. When the function $f(x, y)$ is nonnegative, we interpret its integral over a region D in the xy-plane as the volume of the solid that lies under the surface $z = f(x, y)$ and over D. The plot on the left below shows the surface $z = 3/(x^2 + y^2 + 1)$ over the square described by $-2 \le x \le 2$ and $-2 \le y \le 2$. The plot on the right depicts a volume approximation obtained by evaluating f at the center of each rectangle of a 12 by 12 grid.

```
f[x_, y_] := 3 / (x^2 + y^2 + 1)
GraphicsRow[
  {Plot3D[f[x, y], {x, -2, 2}, {y, -2, 2}, BoxRatios → Automatic],
   With[{h = 4 / 12}, Graphics3D[{(Cuboid @@ # &) /@ Flatten[
       Table[{{x, y, 0}, {x + h, y + h, f[x + h / 2, y + h / 2]}},
         {x, -2, 2 - h, h}, {y, -2, 2 - h, h}], 1]}, Axes → True]]}]
```

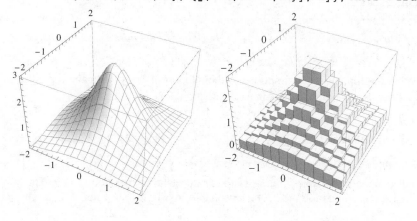

The following is a function that computes such a "midpoint approximation" to the integral of an expression $\xi = f(x, y)$ over a rectangular region described by $a \le x \le b$ and $c \le y \le d$. The parameters m and n are the number of subdivisions in the x and y directions, respectively. Note that each rectangle formed by the resulting grid has area $((b - a)/m)((d - c)/n)$.

```
midPt2D[ξ_, {x_, a_, b_, m_}, {y_, c_, d_, n_}] :=
```
$$\frac{(b-a)\ (d-c)}{m\ n} \sum_{i=1}^{m} \sum_{j=1}^{n} \left(\xi\ /.\ \left\{ x \to a + \frac{(b-a)\ (i-.5)}{m},\ y \to c + \frac{(d-c)\ (j-.5)}{n} \right\} \right)$$

The following creates a `Manipulate` panel with which you can view volume approximations corresponding to 1, 2, 4, 8, 16, and 32 subdivisions in each direction.

```
surf = Plot3D[f[x,y],{x,-2,2},{y,-2,2},Mesh→{30,30},PlotStyle→None];
Manipulate[h = 4 / n;
  Show[surf, Graphics3D[{EdgeForm[Gray], (Cuboid@@#&) /@
      Flatten[Table[{{x, y, 0}, {x+h, y+h, f[x+h/2, y+h/2]}},
        {x, -2, 2-h, h}, {y, -2, 2-h, h}], 1]}], BoxRatios → Automatic],
  {n, 2^Range[0, 5], ControlType → SetterBar}]
```

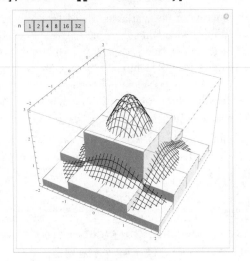

- ## Example 5.1.1

  By computing several midpoint approximations, estimate the volume of the solid that lies under the surface $f(x, y) = 3/(x^2 + y^2 + 1)$ and above the square described by $-2 \le x \le 2$ and $-2 \le y \le 2$ to two decimal places.

This computes midpoint approximations to the integral of $f(x, y)$ over the region of interest, using 8, 16, 32, and 64 subdivisions in each direction:

```
f[x_, y_] := 3 / (x² + y² + 1);
Table[midPt2D[f[x, y], {x, -2, 2, 2ⁿ}, {y, -2, 2, 2ⁿ}], {n, 3, 7}]
```
> {16.9433, 16.9025, 16.8922, 16.8897, 16.889}

The numbers suggest that the volume of the region is approximately 16.89. The surface is shown in the preceeding plot. Is 16.89 a reasonable answer for the volume under the surface?

- ## Example 5.1.2

    By computing several midpoint approximations, estimate the integral of

    $f(x, y) = x y^2 \cos x + x \sin y$ *over the rectangle described by* $1 \le x \le 2$ *and* $0 \le y \le 3$ *to two decimal places.*

    Let's begin by computing the midpoint approximations with four subdivisions in the x direction and twelve in the y direction.

    ```
    f[x_, y_] := x y^2 Cos[x] + x Sin[y];
    midPt2D[f[x, y] , {x, 1, 2, 4}, {y, 0, 3, 12}]
    ```

    3.22393

    Now we'll compute a few more approximations by successively doubling the number of subdivisions in each direction.

    ```
    Table[midPt2D[f[x, y], {x, 1, 2, 2ⁿ 4}, {y, 0, 3, 2ⁿ 12}], {n, 1, 4}]
    ```

    {3.18426, 3.17436, 3.17189, 3.17127}

    These numbers suggest that to two decimal places the integral is 3.17.

- # Iterated Integrals

- ## Example 5.1.3

    Let's consider the integral of the function

    ```
    f[x_, y_] := x³ y² + x y
    ```

    over the rectangle where $0 \le x \le 1$ and $1 \le y \le 2$. A sequence of midpoint approximations,

    ```
    Table[midPt2D[f[x, y], {x, 0, 1, 2ⁿ}, {y, 1, 2, 2ⁿ}], {n, 3, 6}]
    ```

    {1.32845, 1.33211, 1.33303, 1.33326}

    suggests that perhaps the exact value of the integral is $4/3$. Let's see what happens if we integrate $f(x, y)$ with respect to x over $0 \le x \le 1$ while treating y as a constant.

    $$\int_0^1 \left(x^3 y^2 + x y \right) \, dx$$

    $$\frac{y}{2} + \frac{y^2}{4}$$

    Note that the result is a function of y only. Now let's integrate that function over $1 \le y \le 2$:

    $$\int_1^2 \% \, dy$$

    $$\frac{4}{3}$$

    Sure enough, the result coincides with our conjecture about the value of the double integral. Note that the two steps can be combined as follows:

$$\int_1^2 \left(\int_0^1 \left(x^3\, y^2 + x\, y \right)\, dx \right) dy$$

$$\frac{4}{3}$$

We should wonder whether the same value would be obtained if we integrate first with respect to y and then with respect to x. The result of integrating over $1 \le y \le 2$ is the following function of x:

$$\int_1^2 \left(x^3\, y^2 + x\, y \right)\, dy$$

$$\frac{3\,x}{2} + \frac{7\,x^3}{3}$$

Now integrating the result over $0 \le x \le 1$ gives

$$\int_0^1 \%\, dx$$

$$\frac{4}{3}$$

Again, note that the two steps can be combined as follows:

$$\int_0^1 \left(\int_1^2 \left(x^3\, y^2 + x\, y \right)\, dy \right) dx$$

$$\frac{4}{3}$$

An integral computed in this way is an **iterated integral**.

■ **Example 5.1.4**

Compute, as an iterated integral, the integral of $f(x, y) = y \cos x + x \sin y$ over the rectangle given by $0 \le x \le \pi/2$ and $0 \le y \le \pi$. Verify that the result is the same when the order of integration is reversed.

Integrating first with respect to x and then with respect to y, we find

$$\int_0^{\pi/2} \left(y\, \text{Cos}[x] + x\, \text{Sin}[y] \right)\, dx$$

$$\int_0^{\pi} \%\, dy$$

$$y + \frac{1}{8}\, \pi^2\, \text{Sin}[y]$$

$$\frac{3\,\pi^2}{4}$$

Integrating first with respect to y and then with respect to x, we find

$$\int_0^\pi (\mathbf{y\ Cos[x]\ +\ x\ Sin[y]})\ \mathbf{dy}$$

$$\int_0^{\pi/2} \mathbf{\%\ dx}$$

$$2\ \mathbf{x}\ +\ \frac{1}{2}\ \pi^2\ \mathbf{Cos[x]}$$

$$\frac{3\ \pi^2}{4}$$

■ General Regions

Suppose that f is continuous over a region D in the plane described by

$$a \le x \le b \text{ and } g(x) \le y \le h(x).$$

Then the integral of f over D is given by the iterated integral

$$\int_a^b \int_{g(x)}^{h(x)} f(x, y)\, dy\, dx.$$

If D is described instead by

$$g(y) \le x \le h(y) \text{ and } c \le y \le d,$$

then the integral of f over D is given by

$$\int_c^d \int_{g(y)}^{h(y)} f(x, y)\, dx\, dy.$$

■ Example 5.1.5

Find the exact volume of the solid that lies under the surface $z = x\,y^2$ and over the triangular region described by $0 \le x \le 1$ and $x \le y \le 2x$.

Because of the way the region is described, we must integrate first with respect to y and then x. The volume is

$$\int_0^1 \int_x^{2x} \mathbf{x\ y^2\ dy\ dx}$$

$$\frac{7}{15}$$

The following produces a picture of the solid (not to scale). Notice the use of the `Region-Function` and `Filling` options.

```
Plot3D[x y², {x, 0, 1}, {y, 0, 2}, BoxRatios → {1, 1, 1},
  ViewPoint → {2, -1.5, 1}, RegionFunction →
    Function[{x, y}, x ⩽ y ⩽ 2 x && x ⩽ 1], Filling → Bottom]
```

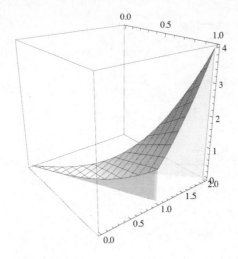

- ## Example 5.1.6

Find the exact volume of the solid that lies under the surface $z = x^2 y^2$ and over the region described by $0 \leq x \leq \sin y$ and $0 \leq y \leq \pi$.

Because of the way the region is described, we must integrate first with respect to y and then x. The volume is

$$\int_0^\pi \int_0^{\text{Sin}[y]} x^2 y^2 \, dx \, dy$$

$$\frac{2}{81} \left(-40 + 9\pi^2 \right)$$

The following produces a picture of the solid (not to scale):

```
Plot3D[x² y², {x, 0, 1}, {y, 0, π},
  BoxRatios → {1, 1, 1}, ViewPoint → {2, -3, 1.0},
  RegionFunction → Function[{x, y}, 0 ≤ x ≤ Sin[y]], Filling → Bottom]
```

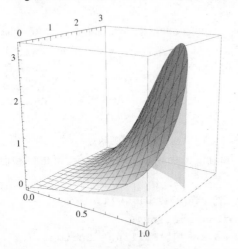

Example 5.1.7

Find the exact volume of the solid that lies under the surface $z = 1 - 4x^2 y^2$ and over the unit disk, i.e., the region inside the unit circle.

This time, let's look at the solid first.

$$\texttt{Plot3D}\left[\texttt{1 - 4 x}^2 \texttt{ y}^2\texttt{, \{x,-1.1,1.1\}, \{y,-1.1,1.1\}, BoxRatios} \rightarrow \texttt{Automatic,}\right.$$
$$\left.\texttt{RegionFunction} \rightarrow \texttt{Function}\left[\texttt{\{x, y\}, x}^2 \texttt{+ y}^2 \leq \texttt{1}\right]\texttt{, Filling} \rightarrow \texttt{Bottom}\right]$$

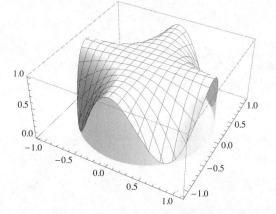

We can take advantage of symmetry by integrating the function over the unit quarter-disk in the first quadrant and then multiplying by four. The unit quarter-disk in the first quadrant can be described by

$$0 \leq x \leq 1 \text{ and } 0 \leq y \leq \sqrt{1 - x^2} \, .$$

The volume of the solid is therefore

$$4 \int_0^1 \int_0^{\sqrt{1-x^2}} \left(1 - 4 x^2 y^2\right) dy\, dx$$

$$\frac{5\pi}{6}$$

5.2 Polar Coordinates

> Double integrals of functions expressed in polar coordinates are discussed in Section 16.4 of Stewart's *Calculus*. (ET:15.4)

Suppose that f is continuous over a region D in the plane described in polar coordinates by

$$g(\theta) \le r \le h(\theta) \text{ and } \alpha \le \theta \le \beta.$$

Then the integral of f over D is given by the iterated integral

$$\int_\alpha^\beta \int_{g(\theta)}^{h(\theta)} f(r\cos\theta, r\sin\theta)\, r\, dr\, d\theta.$$

𝕲 *Note the "extra" factor of r in the integrand.*

■ Example 5.2.1

Use polar coordinates to find integral of $f(x, y) = (x - y)^2/2$ over the sector of the unit disk in which $-\pi/4 \le \theta \le \pi/4$.

The region is described by $-\pi/4 \le \theta \le \pi/4$ and $0 \le r \le 1$. The plots the solid.

```
f[x_, y_] := (x - y)^2 / 2;
Plot3D[f[x, y], {x, -0.02, 1.02}, {y, -0.75, 0.75},
  BoxRatios → Automatic, ViewPoint → {3, 5, 3},
  RegionFunction → Function[{x, y},
    -π / 4 ≤ ArcTan[y / x] ≤ π / 4 && x^2 + y^2 ≤ 1], Filling → Bottom]
```

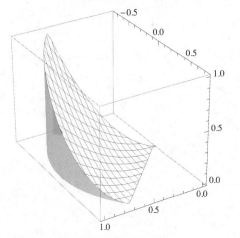

The appropriate computation is

$$\int_{-\pi/4}^{\pi/4} \int_0^1 \texttt{f[r Cos[\theta], r Sin[\theta]] r dr d\theta}$$

$$\frac{\pi}{16}$$

■ Example 5.2.2

Use polar coordinates to compute the volume of the solid that lies under the surface
$z = (x^2 + 1) y$ *and above the disk bounded by the circle* $x^2 + (y-1)^2 = 1$.

We first need a polar description of the circle. Toward that end, let

```
eqn := x^2 + (y - 1)^2 == 1
```

and notice that in polar coordinates this becomes

```
eqn /. {x → r Cos[θ], y → r Sin[θ]} // Simplify
```

$$r^2 == 2 \, r \, \text{Sin}[\theta]$$

Dividing each side by r reveals that $r = 2 \sin \theta$ is the equation of the circle. Consequently, the disk bounded by the circle is described by $0 \le \theta \le \pi$ and $0 \le r \le 2 \sin \theta$. So the integral that gives the volume is computed as follows:

```
f[x_, y_] = (x^2 + 1) y;
```

$$\int_0^\pi \int_0^{2\,\text{Sin}[\theta]} f[r\,\text{Cos}[\theta], \, r\,\text{Sin}[\theta]] \, r \, dr \, d\theta$$

$$\frac{5\,\pi}{4}$$

The following shows a picture of the solid.

```
Plot3D[f[x, y], {x, -1.1, 1.1}, {y, -.1, 2.1},
  BoxRatios → {1, 1, 1}, ViewPoint → {-2, -2, 1},
  RegionFunction → Function[{x, y}, x^2 + (y - 1)^2 ≤ 1],
  Filling → Bottom]
```

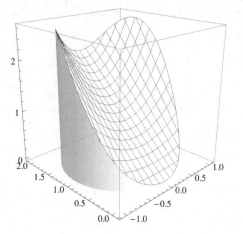

5.3 Applications

Sections 16.5-6 of Stewart's *Calculus* deal with applications of double integrals. (ET:15.5-6)

■ Mass, Moments, and Centers of Mass

For a thin plate—or *lamina*—occupying a region D in the plane and having mass density $\rho(x, y)$ the **moment about the x-axis** is defined as

$$M_x = \iint_D y \, \rho(x, y) \, dA.$$

Similarly, the lamina's **moment about the y-axis** is defined as

$$M_y = \iint_D x \, \rho(x, y) \, dA.$$

The **mass** of the lamina is simply the integral of mass density:

$$m = \iint_D \rho(x, y) \, dA.$$

The **center of mass** of the lamina is the point $(\overline{x}, \overline{y})$, where

$$\overline{x} = M_y / m \ \text{ and } \ \overline{y} = M_x / m.$$

■ Example 5.3.1

A lamina occupies the region in the first quadrant bounded by the graphs of $y = x^2$ and $x = 1$. Its mass density is given by

```
ρ[x_, y_] := 3 x + y²
```

Find the mass and the center of mass of the lamina. Then make a `DensityPlot` *over the region, including the curve $y = x^2$ and a point at the center of mass.*

The region is described by the inequalities $0 \le x \le 1$ and $0 \le y \le x^2$; so the mass is

```
m = ∫₀¹ ∫₀ˣ² ρ[x, y] dy dx
```

$$\frac{67}{84}$$

The moments about the x and y axes, respectively, are

```
xMom = ∫₀¹ ∫₀ˣ² y ρ[x, y] dy dx
```

$$\frac{5}{18}$$

```
yMom = ∫₀¹ ∫₀ˣ² x ρ[x, y] dy dx
```

$$\frac{77}{120}$$

The lamina's center of mass is then

```
ctrMass = {yMom / m, xMom / m}
```

$$\left\{\frac{539}{670}, \frac{70}{201}\right\}$$

```
Show[DensityPlot[ρ[x, y], {x, 0, 1}, {y, 0, 1}, Frame → False,
   ColorFunction → Function[x, GrayLevel[.97 - .8 x]],
    RegionFunction → Function[{x, y}, 0 ≤ y ≤ x²]],
   Graphics[{PointSize[.03], Point[ctrMass]}]]
```

▪ Example 5.3.2

A lamina occupies the region inside the circle $(x - 1)^2 + y^2 = 1$. *Its mass density is proportional to the distance from the origin and given by*

```
ρ[x_, y_] := 1/3 √(x² + y²)
```

Find the mass and the center of mass of the lamina. Make a `DensityPlot` *over the region, including the circle and a point at the center of mass.*

We'll use polar coordinates. The equation of the circle in polar coordinates is $r = 2\cos\theta$; so the region is described by the inequalities $0 \le r \le 2\cos\theta$ and $-\pi/2 \le \theta \le \pi/2$. The mass is therefore

$$\mathbf{m} = \int_{-\pi/2}^{\pi/2}\int_0^{2\,\text{Cos}[\theta]} \rho[r\,\text{Cos}[\theta], r\,\text{Sin}[\theta]]\,r\,dr\,d\theta$$

$$\frac{32}{27}$$

The moments about the *x* and *y* axes, respectively, are

$$\mathbf{xMom} = \int_{-\pi/2}^{\pi/2}\int_0^{2\,\text{Cos}[\theta]} r\,\text{Sin}[\theta]\,\rho[r\,\text{Cos}[\theta], r\,\text{Sin}[\theta]]\,r\,dr\,d\theta$$

0

and

$$\text{yMom} = \int_{-\pi/2}^{\pi/2} \int_{0}^{2\,\text{Cos}[\theta]} \text{r Cos}[\theta] \; \rho[\text{r Cos}[\theta], \text{ r Sin}[\theta]] \text{ r dr d}\theta$$

$$\frac{64}{45}$$

(Note that the calculation of M_x was unnecessary; it is zero by symmetry.) The lamina's center of mass is therefore

```
ctrMass = {yMom / m, xMom / m}
```

$$\left\{\frac{6}{5}, \; 0\right\}$$

Now the plot:

```
Show[DensityPlot[ρ[x, y], {x, 0, 2}, {y, -1, 1}, Frame → False,
   ColorFunction → Function[x, GrayLevel[.97 - .8 x]],
      RegionFunction → Function[{x, y}, (x - 1)² + y² ≤ 1]],
   Graphics[{PointSize[.03], Point[ctrMass]}]]
```

■ Surface Area

Let D be a region in the plane on which the function f is continuous, and let S be the surface $\{(x, y, z) \mid z = f(x, y) \text{ and } (x, y) \in D\}$. The area of S is

$$A(S) = \iint_{D} \sqrt{f_x(x, y)^2 + f_y(x, y)^2 + 1} \; dA.$$

■ Example 5.3.3

Find the area of the part of the surface $z = 1 - 4x^2 y^2$ that lies above the unit disk. (This is the top of the solid in Example 5.1.7.)

First we'll find the partial derivatives.

```
f[x_, y_] := 1 - 4 x² y²;
{fx[x_, y_] = ∂ₓ f[x, y], fy[x_, y_] = ∂_y f[x, y]}
```

$$\left\{-8\,x\,y^2, \; -8\,x^2\,y\right\}$$

It will be convenient to use polar coordinates. Let's have a look at the function we want to integrate (including the "extra r").

$$\text{integrand} = r \sqrt{\text{fx}[r\,\text{Cos}[\theta],\, r\,\text{Sin}[\theta]]^2 + \text{fx}[r\,\text{Cos}[\theta],\, r\,\text{Sin}[\theta]]^2 + 1}$$

$$r \sqrt{1 + 128\, r^6 \text{Cos}[\theta]^2\, \text{Sin}[\theta]^4}$$

Mathematica will not be sucessful in finding the exact value of the integral (try it); so we'll just ask for a numerical approximation with `NIntegrate`.

> 8 **NIntegrate[integrand, {r, 0, 1}, {θ, 0, π / 4}]**

> 4.23151

Is this a reasonable answer? Why?

◆ Exercises

1. Use successively refined midpoint approximations to compute each of the following double integrals to three significant digits. Then use `NIntegrate` to check the result.

   a) $\displaystyle\int_0^{\pi/2}\int_0^{\pi} \sqrt{\cos x + \sin y}\; dx\,dy$ b) $\displaystyle\int_0^2\int_0^3 \frac{dx\,dy}{1 + x^2 + y^2}$ c) $\displaystyle\int_0^1\int_0^1 e^{-(x^2+y^2)}\,dx\,dy$

2. By means of an appropriate iterated integral, compute the integral of $f(x, y) = x^2 + y^2$ over each of the following regions.

   a) the triangle with vertices $(0, 0)$, $(1, 0)$, and $(1, 1)$

   b) the triangle with vertices $(0, 0)$, $(0, 1)$, and $(1, 0)$

   c) the triangle with vertices $(0, 0)$, $(1, 1)$, and $(1, 2)$

   d) the region under the graph of $y = \sin x$, $0 \le x \le \pi$

   e) the unit disk (without using polar coordinates)

   f) the unit disk (using polar coordinates)

3. Compute the volume of the solid that lies under the surface $z = \cos\left(\frac{\pi}{2}(x^2 + y^2)\right)$ and above the unit disk in the xy-plane. Create a plot of the solid (as in Example 5.1.7).

4. Compute the volume of the solid that lies under the surface $z = \sin x \sin y$ and above the square described by $0 \le x \le \pi$, $0 \le y \le \pi$. Plot the surface over the given square.

5. Compute the integral of $f(x, y) = e^{-(x^2+y^2)} \cos(x^2 + y^2)$ over the disk of radius ρ centered at $(0, 0)$. Find the limit of this integral as $\rho \to \infty$.

6. Find the center of mass of the lamina that occupies the region bounded by the unit circle and the line $x + y = 1$, given that its density is $\rho(x, y) = x$.

7. Find the center of mass of the lamina that occupies the region inside the circle $x^2 + (y - 1)^2 = 1$ and outside the circle $x^2 + y^2 = 1$, with $\rho(x, y) = x^2 + y^2$. (Hint: The equation of the first circle in polar coordinates is $r = 2 \sin \theta$.)

8. Compute the area of the portion of the surface $z = x^2 y^2$ that lies above the unit disk.

9. Compute the area of the portion of the surface $z = x^2 y^2$ that lies above the square described by $-1 \le x \le 1$ and $-1 \le y \le 1$.

5.4 Triple Integrals

Triple integrals are defined and developed in Section 16.7 of Stewart's *Calculus*. (ET:15.7)

Suppose that f is continuous on a region E in three dimensional space. If E is described by the inequalities

$$a \leq x \leq b, \ g(x) \leq y \leq h(x), \ \text{and} \ \varphi(x, y) \leq z \leq \psi(x, y),$$

then the integral of f over E is given by the iterated integral

$$\int_a^b \int_{g(x)}^{h(x)} \int_{\varphi(x, y)}^{\psi(x, y)} f(x, y, z) \, dz \, dy \, dx.$$

If E is described instead by

$$a \leq z \leq b, \ g(z) \leq y \leq h(z), \ \text{and} \ \varphi(y, z) \leq x \leq \psi(y, z),$$

then the integral of f over E is given by the iterated integral

$$\int_a^b \int_{g(z)}^{h(z)} \int_{\varphi(y, z)}^{\psi(y, z)} f(x, y, z) \, dx \, dy \, dz.$$

There are four other similar arrangements, which result from the remaining permutations of the variables.

■ Example 5.4.1

Compute the integral of

over the region that lies under the paraboloid $z = x^2 + y^2$ and above the triangle in the xy-plane bounded by the coordinate axes and the line $x + y = 1$.

One way of describing the region over which we'll be integrating is with the inequalities $0 \leq y \leq 1$, $0 \leq x \leq 1 - y$, and $0 \leq z \leq x^2 + y^2$. Here is a plot of it:

```
Plot3D[x² + y², {x, -.1, 1.1}, {y, -.1, 1.1}, BoxRatios → Automatic,
  ViewPoint → {1, -3, 1.5}, Filling → Bottom,
  RegionFunction → Function[{x, y}, 0 ≤ y ≤ 1 && 0 ≤ x ≤ 1 - y]]
```

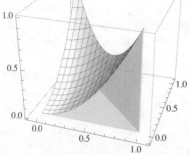

Finally, the integral we want is

$$\int_0^1 \int_0^{1-y} \int_0^{x^2+y^2} \texttt{f[x, y, z] dz dx dy}$$

$$\frac{13\,\pi}{3584}$$

Volume and Mass

The mass of a solid occupying a region E in three dimensional space is the triple integral over E of mass density $\rho(x, y, z)$. When $\rho(x, y, z) = 1$ throughout E, the result is numerically equivalent to the volume of the solid.

Example 5.4.2

A solid occupies the region that lies under the plane $z = x - y/4$ and above the planar region between the graphs of $y = \pm\sin\pi x$ for $0 \le x \le 1$. The mass density is given by

$$\rho\texttt{[x\_, y\_, z\_] := 1 +}\, y^2 + z^2$$

Find the volume of the solid and its total mass.

Inequalities that describe the region are

$$0 \le x \le 1, \; -\sin\pi x \le y \le \sin\pi x, \; \text{and } 0 \le z \le x - y/4.$$

The following is a rendering of the solid.

```
Plot3D[x - y / 4, {x, 0, 1.1}, {y, -1.1, 1.1}, BoxRatios → Automatic,
  ViewPoint → {-3, -2, 1}, Filling → Bottom, RegionFunction →
  Function[{x, y}, -Sin[π x] ≤ y ≤ Sin[π x] && 0 ≤ x ≤ 1] ]
```

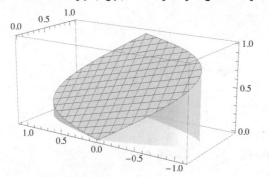

The volume of the solid is

$$\int_0^1 \int_{-\texttt{Sin}[\pi x]}^{\texttt{Sin}[\pi x]} \int_0^{x-y/4} \texttt{dz dy dx}$$

```
% // N
```

$$\frac{2}{\pi}$$

```
0.63662
```

The mass of the solid is

$$\int_0^1 \int_{-\text{Sin}[\pi x]}^{\text{Sin}[\pi x]} \int_0^{x-y/4} \rho[x, y, z] \, dz \, dy \, dx$$

% // N

$$-\frac{4}{\pi^3} + \frac{113}{36 \pi}$$

0.870133

■ Centers of Mass

Given a solid occupying a region E in three dimensional space with mass density $\rho(x, y, z)$, the center of mass of the solid is the point

$$(\overline{x}, \overline{y}, \overline{z}) = \left(\frac{M_{yz}}{m}, \frac{M_{xz}}{m}, \frac{M_{xy}}{m}\right),$$

where M_{yz}, M_{xz}, and M_{xy} are the moments of the solid about the yz-, xz-, and xy-planes, respectively. They are defined by

$$M_{yz} = \iiint_E x\rho(x, y, z) \, dV, \quad M_{xz} = \iiint_E y\rho(x, y, z) \, dV, \quad M_{xy} = \iiint_E z\rho(x, y, z) \, dV.$$

■ Example 5.4.3

Find the center of mass of the solid in Example 5.4.2.

This is the density function:

$$\rho[x\_, y\_, z\_] := 1 + y^2 + z^2$$

The moments about the coordinate planes are

$$\mathbf{myz} = \int_0^1 \int_{-\text{Sin}[\pi x]}^{\text{Sin}[\pi x]} \int_0^{x-y/4} x\rho[x, y, z] \, dz \, dy \, dx$$

$$\frac{32}{\pi^5} - \frac{1466}{81 \pi^3} + \frac{113}{36 \pi}$$

$$\mathbf{mxz} = \int_0^1 \int_{-\text{Sin}[\pi x]}^{\text{Sin}[\pi x]} \int_0^{x-y/4} y\rho[x, y, z] \, dz \, dy \, dx$$

$$\frac{40}{81 \pi^3} - \frac{199}{450 \pi}$$

$$\mathbf{mxy} = \int_0^1 \int_{-\text{Sin}[\pi x]}^{\text{Sin}[\pi x]} \int_0^{x-y/4} z\rho[x, y, z] \, dz \, dy \, dx$$

$$\frac{24}{\pi^5} - \frac{905}{81 \pi^3} + \frac{4333}{2400 \pi}$$

The mass of the solid is

$$m = \int_0^1 \int_{-\text{Sin}[\pi x]}^{\text{Sin}[\pi x]} \int_0^{x-y/4} \rho[x, y, z] \, dz \, dy \, dx$$

$$-\frac{4}{\pi^3} + \frac{113}{36\pi}$$

Thus, the solid's center of mass is

```
{myz, mxz, mxy} / m // Simplify
% // N
```

$$\left\{ \frac{10\,368 - 5864\,\pi^2 + 1017\,\pi^4}{9\,\pi^2\,\left(-144 + 113\,\pi^2\right)}, \ \frac{4000 - 3582\,\pi^2}{225\,\left(-144 + 113\,\pi^2\right)}, \ \frac{1\,555\,200 - 724\,000\,\pi^2 + 116\,991\,\pi^4}{1800\,\pi^2\,\left(-144 + 113\,\pi^2\right)} \right\}$$

```
{0.597604, -0.143469, 0.336462}
```

5.5 Cylindrical and Spherical Coordinates

Triple integrals in cylindrical and spherical coordinates are discussed in Section 16.8 of Stewart's *Calculus*. (ET:15.8)

Recall from Section 2.3 the definitions of cylindrical and spherical coordinates embodied in the following conversion formulas. (You may want to review Section 2.3 at this time.)

```
cylToRect[{r_, θ_, z_}] := {r Cos[θ], r Sin[θ], z}
sphrToRect[{ρ_, θ_, φ_}] := {ρ Sin[φ] Cos[θ], ρ Sin[φ] Sin[θ], ρ Cos[φ]}
```

■ Triple Integrals in Cylindrical Coordinates

If f is continuous on a region E that can be described with cylindrical coordinates by the inequalities

$$\alpha \le \theta \le \beta, \ g(\theta) \le r \le h(\theta), \ \text{and} \ \varphi(r, \theta) \le z \le \psi(r, \theta),$$

then

$$\iiint_E f(x, y, z)\, dV = \int_\alpha^\beta \int_{g(\theta)}^{h(\theta)} \int_{\varphi(r,\theta)}^{\psi(r,\theta)} f(r\cos\theta, r\sin\theta, z)\, r\, dz\, dr\, d\theta.$$

♡ Notice the "extra factor" of r, just as in double integrals in polar coordinates.

■ Example 5.5.1

Use cylindrical coordinates to recompute the volume of the solid in Example 5.1.7, which lies under the surface $z = 1 - 4\,x^2\,y^2$ and over the unit disk in the xy-plane. Then, assuming that the density of the solid is given by

```
ρ[{x_, y_, z_}] := 3 - 2 (x² + y²)
```

find the mass of the solid and its center of mass.

The solid is easily described in cylindrical coordinates by

$$0 \le \theta \le 2\pi, \ 0 \le r \le 1, \ \text{and} \ 0 \le z \le 1 - 4r^4 \cos^2\theta \sin^2\theta.$$

For convenience, let's define

zTop = 1 - 4 r^4 Cos[θ]^2 Sin[θ]^2

1 - 4 r^4 Cos[θ]^2 Sin[θ]^2

So the volume is

$$\int_0^{2\pi} \int_0^1 \int_0^{zTop} \mathbf{1 \, r \, dz \, dr \, d\theta}$$

$\dfrac{5\pi}{6}$

The mass is

$$\mathbf{m} = \int_0^{2\pi} \int_0^1 \int_0^{zTop} \boldsymbol{\rho}\mathbf{[cylToRect[\{r, \theta, z\}]] \, r \, dz \, dr \, d\theta}$$

$\dfrac{7\pi}{4}$

The moments about the coordinate planes are

$$\mathbf{myz} = \int_0^{2\pi} \int_0^1 \int_0^{zTop} \mathbf{r \, Cos[\theta] \, \boldsymbol{\rho}[cylToRect[\{r, \theta, z\}]] \, r \, dz \, dr \, d\theta}$$

0

$$\mathbf{mxz} = \int_0^{2\pi} \int_0^1 \int_0^{zTop} \mathbf{r \, Sin[\theta] \, \boldsymbol{\rho}[cylToRect[\{r, \theta, z\}]] \, r \, dz \, dr \, d\theta}$$

0

$$\mathbf{mxy} = \int_0^{2\pi} \int_0^1 \int_0^{zTop} \mathbf{z \, \boldsymbol{\rho}[cylToRect[\{r, \theta, z\}]] \, r \, dz \, dr \, d\theta}$$

$\dfrac{4\pi}{5}$

(Note that the first two of these calculations were unnecessary because of symmetry.) The center of mass of the solid is therefore

{myz, mxz, mxy} / m

$\left\{0, \ 0, \ \dfrac{16}{35}\right\}$

■ Triple Integrals in Spherical Coordinates

If f is continuous on a region E that can be described with spherical coordinates by

$$\alpha \le \phi \le \beta, \ \gamma \le \theta \le \sigma, \ \text{and} \ g(\theta, \phi) \le \rho \le h(\theta, \phi),$$

then

$$\iiint_E f(x, y, z)\,dV = \int_\alpha^\beta \int_\gamma^\sigma \int_{g(\theta,\phi)}^{h(\theta,\phi)} f(\rho \sin \phi \cos \theta, \ \rho \sin \phi \sin \theta, \ \rho \cos \phi)\, \rho^2 \sin\phi\, d\rho\, d\theta\, d\phi.$$

♡ Notice the "extra" factor $\rho^2 \sin \phi$.

■ Example 5.5.2

The density of a spherical shell occupying the region $2 \le \rho \le 3$ is proportional to the square of the distance from the point $(0, 0, -3)$ and given by

```
ϱ[{x_, y_, z_}] := (x² + y² + (z + 3)²) / 10
```

Find the mass and the center of mass. (Note that here we use a "curly" rho (ϱ) for the density, since we are using the usual rho (ρ) for the radial spherical coordinate.)

The mass is

$$m = \int_0^{2\pi} \int_0^\pi \int_2^3 \varrho[\text{sphrToRect}[\{\rho, \theta, \varphi\}]] \, \rho^2 \, \text{Sin}[\varphi] \, d\rho \, d\varphi \, d\theta$$

$$\frac{992\,\pi}{25}$$

Because of symmetry, each of the moments M_{yz} and M_{xy} will be zero. The moment M_{xy} is

$$mxy = \int_0^{2\pi} \int_0^\pi \int_2^3 \rho \, \text{Cos}[\varphi] \, \varrho[\text{sphrToRect}[\{\rho, \theta, \varphi\}]] \, \rho^2 \, \text{Sin}[\varphi] \, d\rho \, d\varphi \, d\theta$$

$$\frac{844\,\pi}{25}$$

The center of mass of the solid is therefore

```
{0, 0, mxy} / m
```

$$\left\{0, \ 0, \ \frac{211}{248}\right\}$$

◆ Exercises

10. A solid occupies the region between the paraboloid $z = 1 + x^2 + y^2$ and the plane $x + y + z = 0$ that lies above the square in which $0 \le x \le 1$ and $0 \le y \le 1$. Its mass density is $\rho(x, y, z) = 2 + z - x^2 - y^2$. Find the solid's a) volume, b) mass, and c) center of mass.

11. A solid occupies the region bounded by the paraboloid $z = 1 + x^2 + y^2$, the plane $x + y + z = 0$, and the cylinder $x^2 + y^2 = 1$. Its mass density is $\rho(x, y, z) = 1 + z - x^2 - y^2$. Using cylindrical coordinates, find the solid's a) volume, b) mass, and c) center of mass.

A solid occupies the region bounded by the unit sphere $x^2 + y^2 + z^2 = 1$ and the cone $z = x^2 + y^2$. Its mass density is given by $\rho(x, y, z) = 1 + z - x^2 - y^2$. Using spherical coordinates, find the solid's a) volume, b) mass, and c) center of mass.

5.6 Change of Variables

> The change of variables formula for multiple integrals is the subject of Section 16.9 of Stewart's *Calculus*. (ET:15.9)

■ Planar Transformations

Let T be a transformation that takes an ordered pair (u, v) in $\mathbb{R}^2$ to an ordered pair (x, y) in $\mathbb{R}^2$ according to

$$x = g(u, v) \text{ and } y = h(u, v),$$

where g and h are two functions with continuous first-order partial derivatives. We can think of T as a transformation that takes points in the uv-plane to the xy-plane. For a given transformation T and a region S in the uv-plane, we are particularly interested in the region R in the xy-plane that is the **image** of S under T. If we can describe the boundary of the region S in terms of one or more parametric curves, then we can plot the boundary of R by plotting the image of the boundary of S.

■ Example 5.6.1

Let S be the unit square in the uv-plane ($0 \le u \le 1, 0 \le v \le 1$), and let the transformation T be defined by

$$x = 2v^2 \text{ and } y = u^2 + v.$$

Here's how we'll define T in *Mathematica*.

```
trans[{u_, v_}] := {2 v², u² + v}
```

The following describes each of the four edges of S parametrically and then plots them.

```
uvEdges = {{t, 0}, {1, t}, {1 - t, 1}, {0, 1 - t}};
uvRegion = ParametricPlot[Evaluate[uvEdges], {t, 0, 1}]
```

Parametric descriptions of the edges of the image R are found by applying T to those describing the edges of S:

```
xyEdges = Map[trans, uvEdges]
```

$$\{\{0, t^2\}, \{2 t^2, 1 + t\}, \{2, 1 + (1 - t)^2\}, \{2 (1 - t)^2, 1 - t\}\}$$

We can now plot the boundary of R:

```
xyRegion = ParametricPlot[xyEdges, {t, 0, 1}]
```

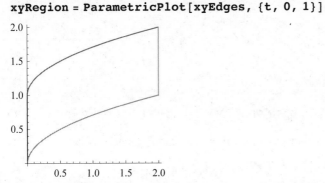

Note that the vertical edges of R are the images of the horizontal edges of S.

■ Example 5.6.2

Let S be the unit circle in the uv-plane, and let the transformation T be the linear transformation defined by

$$x = u + 5v \quad \text{and} \quad y = u - 2v.$$

Plot the boundary of R, the image of S under T.

Here's how we'll define T:

```
trans[{u_, v_}] := {u + 5 v, u - 2 v}
```

The following describes parametrically and plots the boundary of S.

```
uvEdge = {Cos[t], Sin[t]};
uvRegion = ParametricPlot[uvEdge, {t, 0, 2 π}]
```

Parametric description of the boundary of R is found by applying T to the parametric description the boundary of S:

```
xyEdge = trans[uvEdge]
```

$$\{Cos[t] + 5 Sin[t], Cos[t] - 2 Sin[t]\}$$

We can now plot the boundary of *R*, which turns out to be an ellipse:

```
xyRegion = ParametricPlot[xyEdge, {t, 0, 2 π}]
```

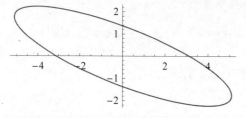

■ Example 5.6.3

Rotations. A transformation of the form

```
rot[θ_][{u_, v_}] := {Cos[θ] u - Sin[θ] v, Sin[θ] u + Cos[θ] v}
```

is a counter-clockwise **rotation** about the origin through the angle θ. Note, for instance, that

```
rot[π / 4][{u, v}] // Factor
```

$$\left\{\left\{\frac{1}{5}\sqrt{2}\left(-5+\sqrt{5}\right),\frac{-5+\sqrt{5}}{5\sqrt{2}}\right\},\left\{\frac{1}{5}\sqrt{2}\left(5+\sqrt{5}\right),\frac{5+\sqrt{5}}{5\sqrt{2}}\right\}\right\}$$

is the counter-clockwise rotation through $\theta = \pi/4$. To illustrate further, let's plot a few rotations of the unit square through multiples of $\pi/9$.

```
uvEdges := {{t, 0}, {1, t}, {1 - t, 1}, {0, 1 - t}};

xyEdges = Flatten[Table[Map[rot[k π / 9], uvEdges], {k, 0, 3}], 1];

ParametricPlot[xyEdges, {t, 0, 1}, Ticks → None]
```

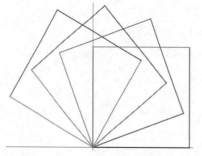

■ The Jacobian Determinant

The **Jacobian matrix** of a transformation *T* given by

$$x = g(u, v) \quad \text{and} \quad y = h(u, v)$$

is defined by

$$\frac{\partial(x, y)}{\partial(u, v)} = \begin{pmatrix} \dfrac{\partial g}{\partial u} & \dfrac{\partial g}{\partial v} \\ \dfrac{\partial h}{\partial u} & \dfrac{\partial h}{\partial v} \end{pmatrix}.$$

The determinant of this matrix,

$$\det \frac{\partial(x, y)}{\partial(u, v)} = \frac{\partial g}{\partial u}\frac{\partial h}{\partial v} - \frac{\partial g}{\partial v}\frac{\partial h}{\partial u},$$

is the **Jacobian determinant** of T—sometimes called simply the *Jacobian* of T. Here are *Mathematica* functions for computing a Jacobian matrix and its determinant.

```
jacobianMat[trans_][u_, v_] := (∂ᵤ trans[[1]]   ∂ᵥ trans[[1]]);
                                 (∂ᵤ trans[[2]]   ∂ᵥ trans[[2]])

jacobianDet[trans_][u_, v_] := Det[jacobianMat[trans][u, v]]
```

In the interest of simplicity, we will adopt the following notation for the Jacobian determinant:

$$\mathcal{J}(u, v) = \det \frac{\partial(x, y)}{\partial(u, v)}.$$

- ## Example 5.6.4

The Jacobian matrix of the transformation Ψ given by

```
trans[{u_, v_}] := {2 v² + 3 u, u² + v}
```

is

```
jacobianMat[trans[{u, v}]][u, v] // MatrixForm
(3    4 v)
(2 u  1  )
```

and the Jacobian determinant is

```
jacobianDet[trans[{u, v}]][u, v]
3 - 8 u v
```

- ## Example 5.6.5

The Jacobian matrix of the rotation T given by

```
rot[θ_][{u_, v_}] := {Cos[θ] u - Sin[θ] v, Sin[θ] u + Cos[θ] v}
```

is

```
jacobianMat[rot[θ][{u, v}]][u, v] // MatrixForm
(Cos[θ]  -Sin[θ])
(Sin[θ]   Cos[θ])
```

and the Jacobian determinant is

```
jacobianDet[rot[θ][{u, v}]][u, v]
% // Simplify
Cos[θ]² + Sin[θ]²
1
```

■ The Change of Variables Formula

If T is a one-to-one, C^1 transformation that maps the region S in the uv-plane onto the region R in the xy-plane, $\mathcal{J}(u, v) \neq 0$ in the interior of S, and f is continuous on R, then

$$\iint_R f(x, y)\, dA = \iint_S f(x(u, v),\, y(u, v))\, |\mathcal{J}(u, v)|\, du\, dv.$$

■ Example 5.6.6

Let T be the transformation described by

$$x = 3u + 2v \ \text{ and } \ y = u^2 - v,$$

and let R be the image in the xy-plane of the unit square S (given by $0 \le u \le 1, 0 \le v \le 1$) in the uv-plane under T. Plot the region R and check that T is one-to-one on S and $\mathcal{J}(u, v) \neq 0$ in the interior of S. Then compute the integral over R of $f(x, y) = x^2 - y$.

First let's define the function and the transformation:

```
f[{x_, y_}] := x² - y; trans[{u_, v_}] := {3 u + 2 v, u² - v}
```

The edges of S are described parametrically (with $0 \le t \le 1$) by

```
edgesS := {{t, 0}, {1, t}, {1 - t, 1}, {0, 1 - t}}
```

The edges of R are then described parametrically by

```
edgesR = Map[trans, edgesS]
```

$$\{\{3\,t, t^2\}, \{3 + 2\,t, 1 - t\}, \{2 + 3\,(1 - t), -1 + (1 - t)^2\}, \{2\,(1 - t), -1 + t\}\}$$

We can now plot R:

```
regionR = ParametricPlot[Evaluate[edgesR], {t, 0, 1}]
```

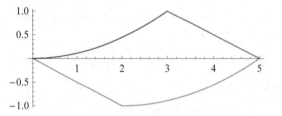

To check that T is one-to-one on S, we will suppose that $(u, v) \in S$ and show that no other point in S has the same image under T. To do that, we'll show that $T(\mu, \nu) = T(u, v)$ has no solution (μ, ν) in S other than $\mu = u, \ \nu = v$.

```
Solve[trans[{μ, ν}] == trans[{u, v}], {μ, ν}]
```

$$\left\{ \{\nu \to v, \mu \to u\}, \left\{ \nu \to \frac{1}{4}\,(9 + 12\,u + 4\,v), \mu \to \frac{1}{2}\,(-3 - 2\,u) \right\} \right\}$$

Note that since $0 \le u \le 1$ and $0 \le v \le 1$, the second solution found does not lie in S. Therefore, T is one-to-one on S.

The Jacobian determinant of T is

```
jacobianDet[trans[{u, v}]][u, v]
```

$-3 - 4 u$

So we see that $\mathcal{J}(u, v) = 0$ only on the vertical line $u = -3/4$, which does not intersect S; therefore $\mathcal{J}(u, v) \neq 0$ in the interior of S. Now finally, we'll compute the integral. By the change of variables formula, and because $|\mathcal{J}(u, v)| = 3 + 4u$ on S, the integral is

$$\int_0^1 \int_0^1 f[\text{trans}[\{u, v\}]]\ (3 + 4 u)\ du\ dv$$

$\dfrac{253}{6}$

■ Example 5.6.7

Let R be the region in the xy-plane that is bounded by the lines $y = x$ and $y = x - 1$ and the circles of radius $\sqrt{2}$ centered at $\left(\sqrt{2}, \sqrt{2}\right)$ and $\left(0, -\sqrt{2}\right)$. Use a change of variables to compute the integral over R of $f(x, y) = x y$.

The region R can be plotted as follows.

```
xyEdges = {{t, t}, {t, t - 1},
    √2 {1 + Cos[π t], 1 + Sin[π t]}, √2 {Cos[π t], -1 + Sin[π t]}};
ParametricPlot[Evaluate[xyEdges], {t, 0, 2},
  PlotRange → {{0, 1.2}, {-.3, .5}}, AxesLabel → {x, y}]
```

Notice that a counter-clockwise rotation through $\theta = \pi/4$ will result in a region with two vertical edges, which will make the integration set-up much simpler.

```
uvEdges = Map[rot[π / 4], xyEdges];
uvRegion = ParametricPlot[Evaluate[uvEdges], {t, 0, 2},
  PlotRange → {{-.2, 1}, {-.1, .9}}, AxesLabel → {u, v}]
```

The centers of the two circles are rotated as follows:

```
Map[rot[π / 4], {{√2 , √2 }, {0, -√2 }}]
```

$$\{\{0, 2\}, \{1, -1\}\}$$

So the resulting circles in the *uv*-plane have the equations

$$u^2 + (v - 2)^2 = 1 \quad \text{and} \quad (u - 1)^2 + (v + 1)^2 = 1.$$

Consequently, the region *S* in the *uv*-plane is described by the inequalities

$$0 \le u \le \sqrt{2}\,/2 \quad \text{and} \quad -1 + \sqrt{1 - (u - 1)^2} \;\le v \le 2 + \sqrt{1 - u^2}\,.$$

The change of variables formula will involve the Jacobian determinant of the *inverse* of this transform-ation, which is just a clockwise rotation through $\pi/4$ radians—that is, a counter-clockwise rotation through $-\pi/4$ radians. The Jacobian determinant of this inverse transformation is

```
jacobianDet[rot[-π / 4] [{u, v}]] [u, v]
```

1

(as it is for any simple rotation). This is the result:

```
f[{x_, y_}] := x y;
```

$$\textbf{Assuming}\left[0 < u < 1, \int_0^{\sqrt{1/2}} \int_{-1+\sqrt{1-(u-1)^2}}^{2+\sqrt{1-u^2}} \textbf{f[rot[-π / 4] [\{u, v\}]] dv du}\right]$$

```
% // N
```

$$\frac{1}{24}\left(19 + 24\sqrt{2} - \sqrt{13 + 16\sqrt{2}} + 6\pi\right)$$

2.74258

■ Example 5.6.8

Find the integral of

```
f[{x_, y_}] := Cos[ (x - y) / (x + y)]
```

over the triangle R in the xy-plane with vertices (0, 0), (1, 1), *and* (2, 0).

The form of $f(x, y)$ suggests that an appropriate change of variables might be

$$u = x - y \quad \text{and} \quad v = x + y.$$

This actually defines the inverse of the transformation we need. So let's define

```
invTrans[{x_, y_}] := {x - y, x + y}
```

and then solve for *x* and *y* :

```
trans[{u_, v_}] =
  {x, y} /. Solve[{u, v} == invTrans[{x, y}], {x, y}][[1]]
```

$$\left\{\frac{u+v}{2}, \frac{1}{2}(-u + v)\right\}$$

The region S in the uv-plane will also be a triangle, since the transformation is linear. To find the vertices of S, we can apply our inverse transformation to the vertices of R.

```
Map[invTrans, {{0, 0}, {1, 1}, {2, 0}}]
```

```
{{0, 0}, {0, 2}, {2, 2}}
```

So S is given by the inequalities $0 \le u \le 2$ and $u \le v \le 2$. The Jacobian determinant of the transformation that maps S onto R is

```
jacobianDet[trans[{u, v}]][u, v]
```

$$\frac{1}{2}$$

Therefore the integral we want is

```
∫₀² ∫ᵤ² f[trans[{u, v}]] (1 / 2) dv du

% // N
```

```
Sin[1]
```

```
0.841471
```

▪ Triple Integrals

Straightforward extension of the change of variables theorem for double integrals results in the following analogous formula for triple integrals:

$$\iiint_R f(x, y, z)\, dV = \iiint_S f(x(u, v, w), y(u, v, w), z(u, v, w))\, |\mathcal{J}(u, v, w)|\, du\, dv\, dw,$$

where $\mathcal{J}(u, v, w)$ is the determinant of the 3×3 Jacobian matrix of the transformation

$$x = x(u, v, w), \quad y = y(u, v, w), \quad \text{and} \quad z = z(u, v, w).$$

The Jacobian determinant can be defined as follows. (To conserve space, we'll now use a single letter Ψ (upper-case psi) to denote the transformation.)

```
jacobianMat3[Ψ_][u_, v_, w_] := ( ∂u Ψ[[1]]  ∂v Ψ[[1]]  ∂w Ψ[[1]] )
                                ( ∂u Ψ[[2]]  ∂v Ψ[[2]]  ∂w Ψ[[2]] ) ;
                                ( ∂u Ψ[[3]]  ∂v Ψ[[3]]  ∂w Ψ[[3]] )

jacobianDet3[Ψ_][u_, v_, w_] := Det[jacobianMat3[Ψ][u, v, w]]
```

▪ Example 5.6.9

Compute the integral of $f(x, y, z) = z\big/\sqrt{1 + x^2 + y^2 + z^2}$ over the top half of the unit sphere ($x^2 + y^2 + z^2 \le 1$, $z \ge 0$) by means of the transformation Ψ given by

$$x = \rho \cos\theta \sin\phi, \quad y = \rho \sin\theta \sin\phi, \quad \text{and} \quad z = \rho \cos\phi.$$

First, we'll enter the function f and transformation Ψ:

```
f[{x_, y_, z_}] := z / √(1 + x² + y² + z²) ;
Ψ[ρ_, θ_, φ_] := {ρ Cos[θ] Sin[φ], ρ Sin[θ] Sin[φ], ρ Cos[φ]}
```

The top half of the unit sphere is the image under Ψ of the region S in $\rho\theta\phi$-space described by

$$0 \le \rho \le 1, \ 0 \le \theta \le 2\pi, \text{ and } 0 \le \phi \le \pi/2.$$

The Jacobian matrix of Ψ is

jacobianMat3[Ψ[ρ, θ, ϕ]][ρ, θ, ϕ] // MatrixForm

$$\begin{pmatrix} \text{Cos}[\theta]\,\text{Sin}[\phi] & -\rho\,\text{Sin}[\theta]\,\text{Sin}[\phi] & \rho\,\text{Cos}[\theta]\,\text{Cos}[\phi] \\ \text{Sin}[\theta]\,\text{Sin}[\phi] & \rho\,\text{Cos}[\theta]\,\text{Sin}[\phi] & \rho\,\text{Cos}[\phi]\,\text{Sin}[\theta] \\ \text{Cos}[\phi] & 0 & -\rho\,\text{Sin}[\phi] \end{pmatrix}$$

and the Jacobian determinant of Ψ is

jacobianDet3[Ψ[ρ, θ, ϕ]][ρ, θ, ϕ] // Simplify

$-\rho^2\,\text{Sin}[\phi]$

whose absolute value on S is simply $\rho^2 \sin\phi$. Therefore, the integral we want is

$$\int_0^{\pi/2} \int_0^{2\pi} \int_0^1 \mathbf{f}[\Psi[\rho, \theta, \phi]]\,\rho^2\,\mathbf{Sin}[\phi]\,\mathbf{d}\rho\,\mathbf{d}\theta\,\mathbf{d}\phi$$

% // N

$-\dfrac{1}{3}\left(-2 + \sqrt{2}\,\right)\pi$

0.613434

♡ *You probably noticed that this example simply involves a conversion to spherical coordinates. The point is that the "extra" $\rho^2 \sin\phi$ factor in the integrand results from the Jacobian determinant of the transformation defined by the conversion to spherical coordinates.*

◆ Exercises

13. Plot the image (in the xy-plane) of both the unit square and the unit disk (in the uv-plane) under each of the following transformations. Also compute the Jacobian determinant of the transformation and describe the set of points in the uv-plane where it is zero.

    a) $x = u + v, \ y = v^2 - u^2$ b) $x = u + v, \ y = uv$

    c) $x = u^2 - v^2, \ y = u^2 + v^2$ d) $x = u^2 - v^2, \ y = 2uv$

14. Let R be the trapezoid in the xy-plane bounded by the lines $y = 1 - x$, $y = 2 - x$, $y = -3x$, and $y = x/2$. Simplify the computation of the integral over R of $f(x, y) = xy$ with a change of variables that rotates R through a clockwise angle of $\pi/4$ radians.

15. Let S be the unit quarter-disk in the first quadrant of the uv-plane, and let R be the image of S under the transformation

$$x = u^2 - v^2, \ y = 2uv.$$

Plot R and compute the integral of the function $f(x, y) = \left(x^2 + y^2\right)^{-1/2}$ over R.

16. Let R be the square in the xy-plane bounded by the lines $x - y = 0$, $x + y = 0$, $x - y = 1$, and $x + y = 1$. Compute the integral of

$$f(x, y) = \sqrt{x + y} \ \cos \sqrt{x - y}$$

over R by means of the change of variables

$$u = \sqrt{x + y}, \quad v = \sqrt{x - y}.$$

17. Let R be the region in the xy-plane bounded by $y = \sqrt{x}$, $y = 0$, and $y = 2\sqrt{x} - 1$. Compute the integral of

$$f(x, y) = \cos\left(\sqrt{x} - y\right)$$

over R by means of the change of variables

$$u = \sqrt{x}, \quad v = \sqrt{x} - y.$$

18. Let R be the region in the xy-plane bounded by $y = x/2$, $y = x$, $xy = 1$, and $xy = 2$. Compute the integral of

$$f(x, y) = e^{xy}$$

over R by means of the change of variables

$$u = y/x, \quad v = xy.$$

19. Let R be the rectangle in the xy-plane bounded by the lines $x - y = 0$, $x + y = 0$, $x - y = \pi/2$, and $x + y = \pi$. Compute the integral of

$$f(x, y) = \cos\left(x^2 - y^2\right)$$

over R by means of the change of variables

$$u = x - y, \quad v = x + y.$$

20. Let R be the planar region inside of the ellipse described by $x^2/9 + y^2 = 1$. Compute the integral over R of

$$f(x, y) = x^2 y$$

by using the change of variables $u = x/3$, $v = y$ and then converting to polar coordinates.

21. Let R be the three-dimensional region inside of the ellipsoid described by $x^2 + y^2/4 + z^2/9 = 1$. Compute the integral over R of

$$f(x, y, z) = x + y + z$$

by using the change of variables $u = x$, $v = y/2$, $w = z/3$ and then converting to spherical coordinates.

6 Vector Calculus

A vector field is a function that maps points in $\mathbb{R}^n$ to points in $\mathbb{R}^n$. We often visualize such a mapping as one that associates vectors with points in space. The velocity of a fluid and the force of a magnetic field, among other phenomena, are vector fields. Vector calculus is the study of vector fields and centers around three higher-dimensional versions of the Fundamental theorem of Calculus, which are known as Green's theorem, Stokes's theorem, and the divergence theorem. Vector calculus is the subject of Chapter 18 of Stewart's *Calculus*.

6.1 Vector Fields

> Vector fields are defined and discussed in Section 18.1 of Stewart's *Calculus*. (ET:17.1)

■ VectorFieldPlot and Its Options

The `VectorFieldPlots` package contains the command `VectorFieldPlot`, which plots two-dimensional vector fields. We load the package as follows.

```
<< VectorFieldPlots`
```

With no specified options, the vector field $\mathbf{F}(x, y) = \langle x - y, x \rangle$ is plotted by `VectorField-Plot` on the rectangle $-2 \le x \le 2$, $-2 \le y \le 2$ as follows.

```
VectorFieldPlot[{x - y, x}, {x, -2, 2}, {y, -1, 1}]
```

The default settings for the `Axes`, `Frame`, and `AspectRatio` options are `False`, `False`, and `Automatic`, respectively. The default for `PlotPoints` is 15 points (vectors) in each direction. In a plot such as the one above, it is probably advantageous to set `PlotPoints` so that fewer points are used in the vertical direction. (We'll also set `Axes→True`.)

```
VectorFieldPlot[{x - y, x}, {x, -2, 2},
  {y, -1, 1}, PlotPoints → {15, 10}, Axes → True]
```

An important option for `VectorFieldPlot` is `ScaleFactor`, which determines the length of the longest vector drawn. Setting `ScaleFactor→None` causes each vector to be drawn with its actual length.

```
VectorFieldPlot[{x - y, x}, {x, -2, 2}, {y, -1, 1},
  PlotPoints → {15, 7}, ScaleFactor → None, Axes → True]
```

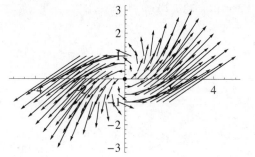

Usually, at least some scaling is required to get a nice plot. The following is a good compromise between `ScaleFactor→None` and `ScaleFactor→Automatic` (the default) for this particular example.

```
VectorFieldPlot[{x - y, x}, {x, -2, 2}, {y, -1, 1},
  PlotPoints → {15, 7}, ScaleFactor → .7, Axes → True]
```

Suppose that we were only concerned with the direction of the vectors and not their lengths. It would be nice to have a plot in which all the vectors had the same length. One way to do this is to divide each vector by its length so that the resulting vectors all have length 1.

$$\texttt{VectorFieldPlot}\Big[\{x - y,\ x\}\ \Big/\ \sqrt{(x-y)^2 + x^2}\ ,$$

$$\{x,\ -2,\ 2\},\ \{y,\ -1,\ 1\},\ \texttt{PlotPoints} \to \{15,\ 7\},\ \texttt{Axes} \to \texttt{True}\Big]$$

Another way to accomplish the same thing is to use the `ScaleFunction` option. This is a little tricky, since `ScaleFunction` must be given as a a *pure function*. The following applies the constant function 1 to the length of each vector.

```
VectorFieldPlot[{x - y, x},
 {x, -2, 2}, {y, -1, 1}, PlotPoints → {15, 7},
 ScaleFunction → Function[x, 1], Axes → True]
```

Another way to specify the `ScaleFunction` is to use the pound sign # to represent the variable and an ampersand & to mark the end of the formula. For instance, in the preceding command, the option `ScaleFunction → Function[x,1]` is equivalent to `Scale-Function → (1&)`. Note the parenthese around 1&; that's important.

To illustrate further, lets apply the square root function to the length of each vector by means of the pure function $\sqrt{\#}$ & .

```
VectorFieldPlot[{x - y, x}, {x, -2, 2}, {y, -1, 1},
 PlotPoints → {15, 7}, ScaleFunction → (√# &), Axes → True]
```

■ **VectorFieldPlot3D**

The `VectorFieldPlots` package also defines the function `VectorFieldPlot3D` for plotting three-dimensional vector fields. This shows what `VectorFieldPlot3D` does with no specified options:

`VectorFieldPlot3D[{y, -x, z / 2}, {x, -1, 1}, {y, -1, 1}, {z, 0, 2}]`

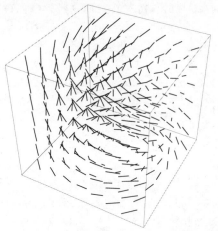

The default value of `PlotPoints` is 7, and the option `VectorHeads` has a default value of `False`. In the next plot, we set `VectorHeads→True`.

`VectorFieldPlot3D[{y, -x, z}, {x, -1, 1},`
  `{y, -1, 1}, {z, 0, 2}, VectorHeads → True]`

The `ScaleFactor` and `ScaleFunction` options of `VectorFieldPlot` can also be used with `VectorFieldPlot3D`.

■ **Conservative (Gradient) Vector Fields**

If $\mathbf{F}(x, y) = \langle p(x, y), q(x, y) \rangle$ is the gradient of some scalar function $f(x, y)$ on an open, connected region D in the plane, the we say that $\mathbf{F}$ is a **conservative vector field** on D and f is a potential function for $\mathbf{F}$ on D.

The `VectorFieldPlots` package also defines `GradientFieldPlot`, which plots the vector field given by the gradient of a function of two variables. The following shows plots of four conservative vector fields.

```
fns = {(1 + x² + y²)^(-1/2), x + y, Cos[π x] Cos[π y], Cos[π x] + Cos[π y]};
GraphicsGrid[Partition[Table[
  GradientFieldPlot[fns[[i]], {x, -1, 1}, {y, -1, 1}], {i, 4}], 2]]
```

The `VectorFieldPlots` package also defines `GradientFieldPlot3D` for plotting vector fields given by the gradient of a function of three variables. The following shows plots of four three-dimensional conservative vector fields.

```
fns = {(1 + x² + y² + z²)^(-1/2), x + y + z,
  Cos[π x] Cos[π y] Cos[π z], Cos[π x] + Cos[π y] + Cos[π z]};
GraphicsGrid[Partition[Table[GradientFieldPlot3D[fns[[i]],
  {x, -1, 1}, {y, -1, 1}, {z, -1, 1}], {i, 4}], 2]]
```

■ Finding a Potential Function

We know that from Section 13.3 of Stewart's *Calculus* that a vector field

$$\mathbf{F}(x, y) = \langle p(x, y), q(x, y) \rangle$$

with continuous first partial derivatives is a **conservative vector field** on an open, connected region D in the plane, if and only if

$$\frac{\partial p}{\partial y} = \frac{\partial q}{\partial x} \text{ for all } (x, y) \text{ in } D.$$

Suppose that we know $\mathbf{F}(x, y) = \langle p(x, y), q(x, y) \rangle$ is a conservative vector field on an open, connected region D in the plane. How might we find a potential function f?

If there is some point (a, b) from which *every* point (x, y) in D can be reached along a path consisting of a horizontal segment followed by a vertical segment, each of which lie in D, then a potential function is given by the formula

$$f(x, y) = \int_a^x p(u, y)\, du + \int_b^y q(a, u)\, du.$$

This formula is valid whenever D is an open rectangle, an open disk, or $\mathbb{R}^2$, for example.

■ Example 6.1.1

Consider the vector field

```
F[{x_, y_}] := {p[x, y], q[x, y]}
```

where

```
p[x_, y_] := 1 + 2 x y - y Sin[x y];
q[x_, y_] := x^2 - x Sin[x y]
```

Since the partial derivatives

```
∂_y p[x, y]
∂_x q[x, y]
2 x - x y Cos[x y] - Sin[x y]
2 x - x y Cos[x y] - Sin[x y]
```

are equal for all points (x, y), we know that $\mathbf{F}$ has a potential function on $D = \mathbb{R}^2$, which certainly has the property described above, where we can take the point (a, b) to be the origin. So a potential function for $\mathbf{F}$ is given by

```
f[x_, y_] = ∫₀ˣ p[u, y] du + ∫₀ʸ q[0, u] du
```

```
-1 + x + x^2 y + Cos[x y]
```

Any function that differs from this one by a constant will also be a potential function for $\mathbf{F}$— in particular, $f(x, y) = xy + x^2 y + \cos xy$.

◆ Exercises

In Exercises 1-5, a) plot the vector field, b) determine whether the vector field is conservative, and *if* the vector field is conservative, c) find a potential function.

1. $\mathbf{F}(x, y) = \langle x + y, \, x - y \rangle$

2. $\mathbf{F}(x, y) = \langle x - y, \, x + y \rangle$

3. $\mathbf{F}(x, y) = \langle \cos(x + y) - x \sin(x + y), \, -x \sin(x + y) \rangle$

4. $\mathbf{F}(x, y) = \left\langle \dfrac{-y}{x^2 + y^2}, \, \dfrac{x}{x^2 + y^2} \right\rangle$ on $D = \{(x, y) \,|\, y > 0\}$

5. $\mathbf{F}(x, y) = \langle x - x\,y, \, -y + x\,y \rangle$

A three-dimensional vector field $\mathbf{F}(x, y, z) = \langle p(x, y, z), q(x, y, z), r(x, y, z) \rangle$, with continuous first partial derivatives, is a conservative vector field on an open, connected region E in three-space, if and only if

$$\frac{\partial p}{\partial y} = \frac{\partial q}{\partial x}, \; \frac{\partial p}{\partial z} = \frac{\partial r}{\partial x}, \text{ and } \frac{\partial q}{\partial z} = \frac{\partial r}{\partial y} \text{ for all } (x, y, z) \text{ in } E.$$

If $\mathbf{F}$ has a potential function, and if there is some point (a, b, c) from which *every* point (x, y, z) in E can be reached along a path consisting of a segment parallel to the x-axis, followed by segment parallel to the y-axis, followed by a segment parallel to the z-axis, each of which lie in E, then a potential function is given by the formula

$$f(x, y, z) = \int_a^x p(u, y, z)\,du + \int_b^y q(a, u, z)\,du + \int_c^z r(a, b, u)\,du.$$

This formula is valid whenever E is an open box, an open ball, or all of $\mathbb{R}^3$, for example.

In Exercises 6-10, a) plot the vector field, b) determine whether the vector field is conservative, and *if* the vector field is conservative, c) find a potential function.

6. $\mathbf{F}(x, y, z) = \langle x + y - z, \, x - y, \, z - x \rangle$

7. $\mathbf{F}(x, y, z) = \langle x - y + z, \, x - y, \, z + x \rangle$

8. $\mathbf{F}(x, y, z) = \langle z, \, 2\,y, \, x + 2\,z \rangle$

9. $\mathbf{F}(x, y, z) = \dfrac{\langle x, y, z \rangle}{\sqrt{x^2 + y^2 + z^2}}$ on $D = \left\{ (x, y, z) \,\big|\, x^2 + y^2 + z^2 \neq 0 \right\}$

10. $\mathbf{F}(x, y, z) = \langle x - x\,y - x\,z, \, -y + x\,y - y\,z, \, -z + x\,y + x\,z \rangle$

6.2 Line Integrals

Line integrals are the subject of Sections 17.2 and 17.3 of Stewart's *Calculus*. (ET:16.2-3)

■ Line Integrals of Scalar Functions

Given a smooth curve C parametrized by $\mathbf{r}(t)$ with $a \le t \le b$, and a continuous function f defined on C, the **line integral of** f **along** C is

$$\int_C f \, ds = \int_a^b f(\mathbf{r}(t)) \, \sqrt{\mathbf{r}'(t) \cdot \mathbf{r}'(t)} \, dt.$$

Note that since we have expressed this formula in a "vector form," it applies to both two- and three-dimensional problems. In particular, $f = f(x, y)$ and $\mathbf{r}(t) = \langle x(t), y(t) \rangle$ when C is a curve in the xy-plane, and $f = f(x, y, z)$ and $\mathbf{r}(t) = \langle x(t), y(t), z(t) \rangle$ when C is a curve in three dimensional space.

■ Example 6.2.1

Compute the line integral of

```
f[{x_, y_}] := x y
             ─────────
             1 + x + 2 y
```

along the unit quarter-circle in the first quadrant from $(1, 0)$ *to* $(0, 1)$. *Use two different parametrizations and check that the same value of the integral is obtained for each.*

Let's first use the standard parametrization

```
r[t_] := {Cos[t], Sin[t]}
```

with $0 \le t \le \pi/2$. The line integral will be the integral over $0 \le t \le \pi/2$ of the function

```
ξ = f[r[t]] √r'[t].r'[t]  // Simplify
```
$$\frac{\mathrm{Cos}[t]\,\mathrm{Sin}[t]}{1 + \mathrm{Cos}[t] + 2\,\mathrm{Sin}[t]}$$

which is

```
  π/2
∫     ξ ⅆt // Simplify
  0

% // N
```
$$\frac{1}{25}\,(15 - 2\,\pi - 6\,\mathrm{Log}[3] + \mathrm{Log}[8])$$

```
0.168183
```

Another parametrization of the same curve (with the same orientation) is

```
r[t_] := {t, √(1 - t²)}
```

with $0 \le t \le 1$. The line integral will be the integral over $0 \le t \le 1$ of the function

$$\xi = f[r[t]] \sqrt{r'[t].r'[t]} \; // \; \textbf{Simplify}$$

$$\frac{t \sqrt{\dfrac{1}{1-t^2}} \sqrt{1-t^2}}{1+t+2\sqrt{1-t^2}}$$

which is

$$\int_0^1 \xi \, dt \; // \; \textbf{Simplify}$$

$$\% \; // \; \textbf{N}$$

$$\frac{1}{25} \left(15 - 2\,\pi + \text{Log}[8] - 2\,\text{Log}[27] \right)$$

0.168183

So we see that the same value of the integral was obtained for each of these two parametrizations of the curve.

■ Example 6.2.2

Compute the line integral of

$$f[\{x\_, y\_\}] := \text{Cos}[x\,y];$$

along the curve $y = x^3$ *from* $(0, 0)$ *to* $(2, 8)$.

Let's use the simple parametrization

$$r[t\_] := \{t, t^3\}$$

with $0 \le t \le 2$. The line integral will be the integral over $0 \le t \le 2$ of

$$\xi = f[r[t]] \sqrt{r'[t].r'[t]} \; // \; \textbf{Simplify}$$

$$\sqrt{1+9\,t^4} \; \text{Cos}[t^4]$$

which is

$$\int_0^2 \xi \, dt \; // \; \textbf{Simplify}$$

$$\% \; // \; \textbf{N}$$

$$\int_0^2 \sqrt{1+9\,t^4} \; \text{Cos}[t^4] \, dt$$

0.775649

Note that *Mathematica* was not able to compute the integral symbolically but did produce a numerical value.

- ## Example 6.2.3

Compute the line integral of

$$f[\{x\_, y\_, z\_\}] := (x^2 + y^2 + z^2)^{-1}$$

along the portion of the helix parametrized by

$$r[t\_] := \{Cos[2 \pi t], Sin[2 \pi t], t/2\}$$

where $0 \leq t \leq 2$.

The line integral will be the integral over $0 \leq t \leq 2$ of the function

$$\xi = f[r[t]] \sqrt{r'[t].r'[t]} \ // \ Simplify$$

$$\frac{2 \sqrt{1+16 \pi^2}}{4 + t^2}$$

which is

$$\int_0^2 \xi \, dt$$

$$\% \ // \ N$$

$$\frac{1}{4} \pi \sqrt{1 + 16 \pi^2}$$

$$9.90081$$

- ## Line Integrals of Vector Fields

Given a smooth curve C parametrized by $\mathbf{r}(t)$ with $a \leq t \leq b$, and a continuous vector field $\mathbf{F}$ defined on C, the **line integral of F along** C is

$$\int_C \mathbf{F} \cdot d\mathbf{r} = \int_a^b \mathbf{F}(\mathbf{r}(t)) \cdot \mathbf{r}'(t) \, dt.$$

Note again that since we have expressed this formula in a "vector form," it applies to both two- and three-dimensional problems. Also, since

$$\int_C \mathbf{F}(\mathbf{r}(t)) \cdot \mathbf{r}'(t) \, dt = \int_C \mathbf{F}(\mathbf{r}(t)) \cdot \mathbf{T}(t) \sqrt{\mathbf{r}'(t) \cdot \mathbf{r}'(t)} \ dt = \int_C \mathbf{F}(\mathbf{r}(t)) \cdot \mathbf{T}(t) \, ds,$$

where $\mathbf{T}(t) = \mathbf{r}'(t) / \sqrt{\mathbf{r}'(t) \cdot \mathbf{r}'(t)}$, and since $\mathbf{F} \cdot \mathbf{T} = \text{comp}_\mathbf{T} \mathbf{F}$ (see Section 1.3), the line integral of the vector field $\mathbf{F}$ along C is the line integral of the scalar function given by the *tangential component* of $\mathbf{F}$ along C.

- **Example 6.2.4**

    Let $\mathbf{F}(x, y) = \langle -y, x \rangle$, *and let* C *be the portion of the parabola* $y = x^4$ *between* $(0, 0)$ *and* $(1, 1)$. *Plot both the vector field and the curve, and compute the integral of* $\mathbf{F}$ *along* C.

    This defines the vector field and the parametrization of the curve:

    ```
    F[{x_, y_}] := {-y, x}
    r[t_] := {t, t⁴}
    ```

    Now here's a plot of the vector field and the curve:

    ```
    Show[VectorFieldPlot[F[{x, y}], {x, -.2, 1.2},
       {y, -.2, 1.1}, PlotPoints → 10, Axes → True, Ticks → None],
       ParametricPlot[r[t], {t, 0, 1}, PlotStyle → Thickness[.006]]]
    ```

    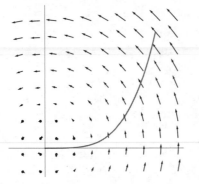

    The line integral of $\mathbf{F}$ along C will be the integral over $0 \le t \le 1$ of the function

    ```
    ξ = F[r[t]].r'[t]
    ```

    $3\,t^4$

    which is

    $$\int_0^1 \xi \; \mathrm{d}t$$

    $$\frac{3}{5}$$

- **Example 6.2.5**

    Let $\mathbf{F}(x, y) = \langle x, -y \rangle$, *and let* C *be the unit quarter-circle between* $(1, 0)$ *and* $(0, 1)$. *Plot both the vector field and the curve, and compute the integral of* $\mathbf{F}$ *along* C.

    This defines the vector field and a parametrization of the curve for $0 \le t \le 1$:

    ```
    F[{x_, y_}] := {x, -y};
    r[t_] := {Cos[t], Sin[t]}
    ```

    A plot of the vector field and the curve is produced by:

    ```
    Show[VectorFieldPlot[F[{x, y}],
       {x, -.1, 1.1}, {y, -.1, 1.1}, PlotPoints → 11],
       ParametricPlot[r[t], {t, 0, π / 2}, Axes → True]
    ```

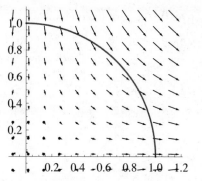

The line integral of **F** along C will be the integral over $0 \le t \le 1$ of the function

```
ξ = F[r[t]].r'[t]
```

$-2 \cos[t] \sin[t]$

which is

$$\int_0^{\pi/2} \xi \, dt$$

-1

■ Example 6.2.6

Let $\mathbf{F}(x, y, z) = \langle -y, x - z, y \rangle$, *and let C be the portion of the helix parameterized by* $\mathbf{r}(t) = (\cos \pi t / 2, \sin \pi t / 2, t)$ *with* $0 \le t \le 1$. *Plot both the vector field and the curve, and compute the integral of* **F** *along C.*

This defines the vector field and a parametrization of the curve for $0 \le t \le 1$:

```
F[{x_, y_, z_}] := {-y, x - z, y};
r[t_] := {Cos[π t / 2], Sin[π t / 2], t}
```

The plot is produced by:

```
Show[ VectorFieldPlot3D[F[{x, y, z}], {x, 0, 1},
   {y, 0, 1}, {z, 0, 1}, PlotPoints → 5, VectorHeads → True],
  ParametricPlot3D[r[t], {t, 0, 1}] ]
```

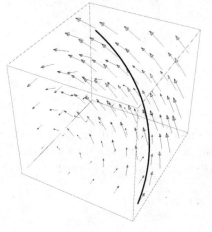

The line integral of **F** along *C* will be the integral over $0 \leq t \leq 1$ of the function

```
ξ = F[r[t]].r'[t] // Simplify
```

$$\frac{1}{2}\left(\pi - \pi t \, \text{Cos}\left[\frac{\pi t}{2}\right] + 2 \, \text{Sin}\left[\frac{\pi t}{2}\right]\right)$$

which is

$$\int_0^1 \xi \, dt$$

```
% // N
```

$$-1 + \frac{4}{\pi} + \frac{\pi}{2}$$

```
1.84404
```

■ Applications

■ Example 6.2.7

A wire of uniform diameter and density is bent in the shape of the curve $y = \cos x$ *where* $-3\pi/2 \leq x \leq 3\pi/2$. *Find its center of mass.*

Let's first parametrize and plot the curve.

```
r[t_] := {t, Cos[t]};
ParametricPlot[r[t], {t, -3 π/2, 3 π/2}]
```

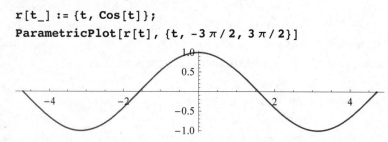

Since the wire's diameter and density are uniform, we may as well assume that its mass density is 1 at every point. Then the mass of the wire will be numerically equivalent to its arc length, which compute as

```
mass = 6 ∫₀^{π/2} √(r'[t].r'[t]) dt
```

$$\text{mass} = 6 \int_0^{\pi/2} \sqrt{r'[t].r'[t]} \, dt$$

```
% // N
```

```
6 EllipticE[-1]
```

```
11.4606
```

By symmetry, the *x*-coordinate of the center of mass will be $\overline{x} = 0$. To find the *y*-coordinate of the center of mass, we first need to compute the moment about the *x*-axis. Since we've assume that the density is 1, the moment about the *x*-axis here is the integral

$$\int_a^b y \, ds = \int_a^b y(t) \sqrt{\mathbf{r}'(t) \cdot \mathbf{r}'(t)} \, dt.$$

After computing this moment, we divide by the mass to obtain $\overline{y}$.

```
xMoment =
```
$$\int_{-3\,\pi/2}^{3\,\pi/2} \mathbf{r[t][[2]]} \ \sqrt{\mathbf{r'[t].r'[t]}} \ dt$$

```
% // N
```

$-\sqrt{2}\ -\ \text{ArcSinh[1]}$

```
-2.29559
```

```
ybar = xMoment / mass // N
```

```
-0.200303
```

So the wire's center of mass is $(0, -.200303)$.

■ Example 6.2.8

A particle travels around the unit square $(0 \le x \le 1, 0 \le y \le 1)$ exactly once, moving counterclockwise along the edges of the square in the presence of a force field given by

$$\mathbf{F[\{x\_,\ y\_\}]\ :=\ \{x - y,\ x^2 + y^2\}}$$

Find the work done on the particle by the force field.

First we'll create a table containing parametrizations of the four edges of the square, each with $0 \le t \le 1$.

```
edges := {{t, 0}, {1, t}, {1 - t, 1}, {0, 1 - t}}
```

(Note that we have taken care to ensure that the orientation of each edge results in counter-clockwise movement.) The following is a plot of the square and the vector field.

```
Show[ParametricPlot[edges, {t, 0, 1}],
  VectorFieldPlot[F[{x, y}], {x, 0, 1}, {y, 0, 1},
    PlotPoints → 8], Axes → None, PlotRange → All]
```

The force field restricted to each of the edges of the square is seen in

```
Map[F, edges]
```

$$\{\{t,\ t^2\},\ \{1 - t,\ 1 + t^2\},\ \{-t,\ 1 + (1 - t)^2\},\ \{-1 + t,\ (1 - t)^2\}\}$$

The functions we will integrate over $0 \leq t \leq 1$ to obtain the work done as the particle moves along the four edges are

```
Table[F[edges[[i]]].∂t edges[[i]], {i, 4}]
```

$$\left\{ t,\ 1 + t^2,\ t,\ -(1-t)^2 \right\}$$

Finally, by integrating $\mathbf{F} \cdot d\mathbf{r}$ along the four edges we obtain the work done as the particle moves along each one, and the sum of those is the total work.

```
∫₀¹ Table[F[edges[[i]]].∂t edges[[i]], {i, 4}] dt
```

```
Apply[Plus, %]
```

$$\left\{ \frac{1}{2},\ \frac{4}{3},\ \frac{1}{2},\ -\frac{1}{3} \right\}$$

2

6.3 Green's Theorem

> Section 17.4 of Stewart's *Calculus*—and proves a special case of—Green's theorem. The notion of a positive orientation of a curve is also defined there. (ET:16.4)

Suppose that D is a region in the plane whose boundary, ∂D, consists of one or more positively-oriented, piecewise-smooth, simple closed curves. Let $\mathbf{F}(x, y) = \langle p(x, y), q(x, y) \rangle$ be a continuously differentiable vector field on an open region containing $D \cup \partial D$. Green's theorem states that

$$\iint_D \left(\frac{\partial q}{\partial x} - \frac{\partial p}{\partial y} \right) dA = \int_{\partial D} \mathbf{F} \cdot d\mathbf{r}.$$

The line integral on the right side is often written as $\int_{\partial D} p\, dx + q\, dy$. (Note that $\mathbf{F} \cdot d\mathbf{r} = \langle p, q \rangle \cdot \langle dx, dy \rangle$.) With that notation, Green's theorem becomes

$$\iint_D \left(\frac{\partial q}{\partial x} - \frac{\partial p}{\partial y} \right) dA = \int_{\partial D} p\, dx + q\, dy.$$

The integrand of the double integral on the left side is sometimes called the **rotation** of $\mathbf{F}$:

$$\mathrm{rot}\, \mathbf{F}(x, y) = \frac{\partial q}{\partial x} - \frac{\partial p}{\partial y}.$$

With that notation, Green's theorem becomes

$$\iint_D \mathrm{rot}\, \mathbf{F}\, dA = \int_{\partial D} \mathbf{F} \cdot d\mathbf{r}.$$

- ## Example 6.3.1

Verify Green's theorem for $\mathbf{F}(x, y) = \langle x\,y, x - y \rangle$ *and* $D \bigcup \partial D$ *the unit disk.*

Let's start by entering

```
F[{x_, y_}] := {p[x, y], q[x, y]}
p[x_, y_] := x y; q[x_, y_] := x - y
```

and computing the rotation of **F**:

```
rot[x_, y_] = ∂x q[x, y] - ∂y p[x, y]
```

1 - x

The double integral $\iint_D \operatorname{rot} F\, dA$—or $\iint_D \left(\dfrac{\partial q}{\partial x} - \dfrac{\partial p}{\partial y} \right) dA$—is

```
∫₀²π ∫₀¹ rot[r Cos[θ], Sin[θ]] r dr dθ
```

π

A parametrization of ∂D (for $0 \le t \le 2\pi$) is

```
r[t_] := {Cos[t], Sin[t]}
```

and the line integral $\int_{\partial D} \mathbf{F} \cdot d\mathbf{r}$ is

```
∫₀²π F[r[t]].r'[t] dt
```

π

- ## Example 6.3.2

Verify Green's theorem if $\mathbf{F}(x, y) = \langle x^2 - y, x\,y \rangle$ *and* D *is the region inside the unit circle and outside the circle of radius* $1/2$ *centered at* $(1/2, 0)$.

Again we start by entering the vector field:

```
F[{x_, y_}] := {p[x, y], q[x, y]}
p[x_, y_] := x² - y; q[x_, y_] := x y
```

and computing its rotation:

```
rot[x_, y_] = ∂x q[x, y] - ∂y p[x, y]
```

1 + y

We'll parametrize the outer boundary of D by

```
outer[t_] := {Cos[t], Sin[t]}
```

for $0 \le t \le 2\pi$ and the inner boundary of D by

```
inner[t_] := {1 / 2, 0} + 1 / 2 {Cos[t], -Sin[t]}
```

for $0 \le t \le 2\pi$, so that each circle is positively oriented with respect to the region D, which amounts to the outer boundary being oriented counterclockwise and the inner boundary being oriented clockwise.

The line integral of F along ∂D is

$$\int_0^{2\pi} \texttt{F[outer[t]].outer'[t] dt} + \int_0^{2\pi} \texttt{F[inner[t]].inner'[t] dt} \; .$$

$$\frac{3\pi}{4}$$

The region inside the outer boundary is described by $0 \le r \le 1$ and $0 \le \theta \le 2\pi$, and the region inside the inner boundary is described by $0 \le r \le \cos\theta$ and $0 \le \theta \le \pi$. So the integral of rot F over D is

$$\int_0^{2\pi}\int_0^1 \texttt{rot[r Cos[}\theta\texttt{], r Sin[}\theta\texttt{]] r dr d}\theta \; -$$

$$\int_0^{\pi}\int_0^{\texttt{Cos[}\theta\texttt{]}} \texttt{rot[r Cos[}\theta\texttt{], r Sin[}\theta\texttt{]] r dr d}\theta$$

$$\frac{3\pi}{4}$$

■ Example 6.3.3

Use Green's theorem to compute the line integral of

$$\mathbf{F}(x,\,y) = \left\langle \cos x^2 - y^2,\, x^3 - \sqrt{y^3+1}\; \right\rangle.$$

around the (positively oriented) unit circle.

We'll begin by entering *p* and *q* and computing rot **F**:

```
p[x_, y_] := Cos[x²] - y²;  q[x_, y_] := x³ - √(y³ + 1) ;
rot[x_, y_] = ∂ₓ q[x, y] - ∂_y p[x, y]
```

$$3\,x^2 + 2\,y$$

According to Green's theorem, we can compute the line integral of **F** around the unit circle by computing the double integral of rot **F** over the unit disk. For this, it will be convenient to use polar coordinates:

$$\int_0^{2\pi}\int_0^1 \texttt{rot[r Cos[}\theta\texttt{], Sin[}\theta\texttt{]] r dr d}\theta$$

$$\frac{3\pi}{4}$$

■ Example 6.3.4

Use Green's theorem to compute the line integral of

$$\mathbf{F}(x,\,y) = \left\langle x\,y - \ln x,\, x^2 + \ln y \right\rangle.$$

around the positively-oriented boundary of: a) the unit square, and b) the region in the first quadrant that lies between the circles of radius 1 and 2 centered at the origin.

Again, we'll begin by entering p and q and computing rot **F**:

```
p[x_, y_] := x y - Log[x]; q[x_, y_] := x² + Log[y];
rot[x_, y_] = ∂x q[x, y] - ∂y p[x, y]
```

x

a) According to Green's theorem, we can compute the line integral of **F** around the boundary of the unit square by computing the double integral of rot **F** over the unit square:

```
⌠¹ ⌠¹
⌡₀ ⌡₀ rot[x, y] dx dy
```

$\dfrac{1}{2}$

b) Again, we need only compute the double integral of rot **F** over the prescribed region D, which in this case is most easily done with the help of polar coordinates:

```
⌠π/2 ⌠²
⌡₀   ⌡₁ rot[r Cos[θ], r Sin[θ]] r dr dθ
```

$\dfrac{7}{3}$

◆ Exercises

11. A circular loop of wire, with radius 1, has mass density given by the square of the distance from one fixed point on the loop. Find the mass of the wire and its center of mass. (Hint: Think of the wire as the unit circle and assume that the fixed point from which distance is measured is $(-1, 0)$.)

12. A wire of uniform density is bent in the shape of the graph of $y = x^2$ for $-1 \le x \le 1$. Find its center of mass.

13. Find the work done on a particle that moves counterclockwise along the unit quarter-circle in the first quadrant through a force field given by $\mathbf{F}(x, y) = \langle x - xy, -y + xy \rangle$.

14. A particle travels around the unit square $(0 \le x \le 1, 0 \le y \le 1)$ exactly once, moving counterclockwise along the edges of the square in the presence of a force field given by $\mathbf{F}(x, y) = \langle xy, x - y \rangle$. Find the work done on the particle by the force field. First do this directly by computing a line integral; then redo the calculation with the help of Green's theorem.

15. A particle travels once around the circle of radius ρ centered at the origin in a counterclockwise direction, while acted upon by a force field given by

$$\mathbf{F}(x, y) = \langle x^2 - y^3 + 6y, \; x^3 - y^2 - 3x \rangle.$$

Is there a value of ρ such that the work done on the particle is zero?

16. A particle moves counterclockwise around the edge of the region bounded by $y = 1 - x^2$ and $y = x^2 - 1$ while being acted upon by a force field given by $\mathbf{F}(x, y) = \langle x^2 - y^3, \; x^3 + y^2 \rangle$. Find the work done on the particle during one trip around the path.

6.4 Surface Integrals

See Sections 17.6-7 of Stewart's *Calculus* for definition and discussion of surface integrals and related concepts. (ET:16.6-7)

■ Surface Integrals of Scalar Functions

Given a smooth surface S parametrized by $\mathbf{r}(u, v)$, $(u, v) \in D$, and a continuous function f defined on S, the **surface integral of f over S** is

$$\iint_S f(x, y, z)\, dS = \iint_S f(\mathbf{r}(u, v))\, \|\mathbf{r}_u \times \mathbf{r}_v\|\, dA.$$

In particular, the surface area of S is

$$\iint_S dS = \iint_D \|\mathbf{r}_u \times \mathbf{r}_v\|\, dA.$$

In the case where S is the graph of $z = g(x, y)$, these formulas become

$$\iint_S f(x, y, z)\, dS = \iint_D f(x, y, g(x, y)) \left(\left(\frac{\partial z}{\partial x}\right)^2 + \left(\frac{\partial z}{\partial y}\right)^2 + 1 \right)^{1/2} dA$$

and

$$\iint_S dS = \iint_D \left(\left(\frac{\partial z}{\partial x}\right)^2 + \left(\frac{\partial z}{\partial y}\right)^2 + 1 \right)^{1/2} dA.$$

Since we will need to compute $\|\mathbf{r}_u \times \mathbf{r}_v\|$ in every such integral over a surface described parametrically, let's go ahead and define a function for doing that:

```
normPerp[r_, {u_, v_}] := With[{c = ∂_u r × ∂_v r}, √c.c ]
```

■ Example 6.4.1

Compute the surface integral over the top half of the unit sphere of

```
f[{x_, y_, z_}] := 1 / (1 + x^2 + y^2)
```

The unit sphere is parametrized with spherical coordinates by

```
r[φ_, θ_] := {Cos[θ] Sin[φ], Sin[θ] Sin[φ], Cos[φ]}
```

and our surface S here corresponds to $0 \le \theta \le 2\pi$ and $0 \le \phi \le \pi/2$. The function of θ and ϕ that we need to integrate is then

```
ξ[φ_, θ_] = f[r[φ, θ]] normPerp[r[φ, θ], {φ, θ}] // Simplify
```

$$-\frac{2\sqrt{\mathrm{Sin}[\phi]^2}}{-3 + \mathrm{Cos}[2\phi]}$$

and the integral is

$$\left\{ \mathtt{int} = \int_0^{\pi/2} \int_0^{2\pi} \xi[\theta, \phi] \, d\theta \, d\phi, \ \mathtt{N[int]} \right\}$$

$$\left\{ \sqrt{2} \ \pi \ \mathtt{ArcCosh}\!\left[\sqrt{2}\right], \ 3.91584 \right\}$$

■ Example 6.4.2

Compute the surface area of the sea shell-like surface parametrized by

```
r[u_, θ_] := {Cos[2 π u] u (1 + Sin[θ]),
   Sin[2 π u] u (1 + Sin[θ]), 4 - u (3 + Cos[θ])}
```

where $0 \le u \le 2$ *and* $0 \le t \le 2\pi$. (See Example 3.6.3 for a plot of the surface.)

The function of u and θ that we need to integrate is

```
ξ[u_, θ_] = normPerp[r[u, θ], {u, θ}] // FullSimplify
```

$$\sqrt{2} \ \sqrt{u^2 \, (1 + \mathtt{Sin}[\theta]) \, \left(5 + 2\,\pi^2\,u^2 + 3\,\mathtt{Cos}[\theta] + 2\,\left(-2 + \pi^2\,u^2\right)\mathtt{Sin}[\theta]\right)}$$

Mathematica is not able to integrate this symbolically, so we will ask for a numerical approximation:

```
NIntegrate[ξ[u, θ], {u, 0, 2}, {θ, 0, 2 π}]
```

```
111.819
```

Look at the plot of the surface in Example 3.6.3 and decide whether you think this is a reasonable answer.

■ Example 6.4.3

Plot the surface $z = (3\,x^2 + 2\,y^2 - 3\,x + y)/4$ *and compute the area of the part of the surface that lies above the unit disk in the xy-plane.*

First let's define

```
g[x_, y_] := (3 x² + 2 y² - 3 x + y) / 4
```

and plot the surface over $-1 \le x \le 1$ and $-1 \le y \le 1$.

```
Plot3D[g[x, y], {x, -1, 1}, {y, -1, 1},
   BoxRatios → Automatic, ViewPoint → {6, 2, 1}, Axes → None]
```

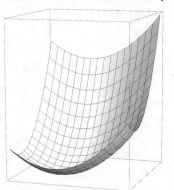

The integrand we need is

$$\xi[\mathtt{x\_, y\_}] = \sqrt{(\partial_x g[x, y])^2 + (\partial_y g[x, y])^2 + 1}$$

$$\sqrt{1 + \frac{1}{16}(-3 + 6x)^2 + \frac{1}{16}(1 + 4y)^2}$$

Since the domain of integration is the unit disk, it will be convenient to use polar coordinates for the integration, and because of the nature of the integrand, we'll settle for a numerical approximation to the integral:

NIntegrate[ξ[r Cos[θ], r Sin[θ]] r, {θ, 0, 2π}, {r, 0, 1}]

4.75158

■ Surface Integrals of Vector Fields

Given a continuous vector field **F** defined on an oriented surface S with unit normal vector **n**, the **surface integral of F over S** is

$$\iint_S \mathbf{F} \cdot d\mathbf{S} = \iint_S \mathbf{F} \cdot \mathbf{n} \, d\mathbf{S}.$$

In other words, the integral of **F** over a surface S is the integral of the *normal component* of **F** over S. (Contrast this with the definition of a line integral of a vector field, which is the integral along a curve of the vector field's *tangential* component.) This integral is also called the **flux** of **F** across S.

If the surface S is parametrized by $\mathbf{r}(u, v)$, $(u, v) \in D$, such that the normal vector $\mathbf{r}_u \times \mathbf{r}_v$ points in the same direction as **n**, then the surface integral of **F** over S is

$$\iint_S \mathbf{F} \cdot d\mathbf{S} = \iint_D \mathbf{F} \cdot (\mathbf{r}_u \times \mathbf{r}_v) \, dA.$$

In the case where S is the graph of $z = g(x, y)$ and **n** points generally upward, this formula becomes

$$\iint_S \mathbf{F} \cdot d\mathbf{S} = \iint_S \mathbf{F} \cdot \left\langle -\frac{\partial z}{\partial x}, -\frac{\partial z}{\partial y}, 1 \right\rangle dA.$$

■ Example 6.4.4

Let $\mathbf{F}(x, y, z) = \langle y^2, z, x^2 \rangle$ *and let S be the portion of the paraboloid $z = 1 - x^2 - y^2$ that lies above the unit disk in the xy-plane. Plot both the vector field and the surface, and compute the flux of **F** across S.*

First let's define the vector field and the function whose graph is the surface:

F[{x_, y_, z_}] := {y^2, z, x^2}; g[x_, y_] := 1 - x^2 - y^2

This is a plot of the vector field and the surface:

```
Needs["VectorFieldPlots`"];
Show[VectorFieldPlot3D[F[{x, y, z}], {x, -1, 1}, {y, -1, 1},
  {z, 0, 1}, PlotPoints → 6, VectorHeads → True, ScaleFactor → .4],
surf = Plot3D[g[x, y], {x, -1, 1}, {y, -1, 1}], PlotRange → {0, 4 / 3}]
```

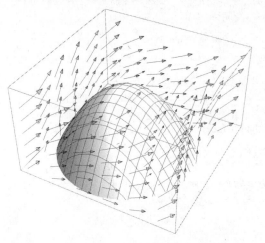

The function we need to integrate over the unit disk is

$$\xi[x\_, y\_] = F[\{x, y, g[x, y]\}] \cdot \{-\partial_x g[x, y], -\partial_y g[x, y], 1\}$$

$$x^2 + 2 \, x \, y^2 + 2 \, y \left(1 - x^2 - y^2\right)$$

Using polar coordinates to set up the integral, we find

$$\int_0^{2\pi} \int_0^1 \xi[r \, \text{Cos}[\theta], r \, \text{Sin}[\theta]] \, r \, dr \, d\theta$$

$$\frac{\pi}{4}$$

■ Example 6.4.5

Let $f(x, y, z) = xyz - x - y - z$ and let S be the part the unit sphere that lies above the first quadrant in the xy-plane. Compute the integral of ∇f over S, and plot the vector field and the surface.

First let's compute the gradient of f using the function grad from Section 4.4.

```
grad[ξ_, χ_?VectorQ] := Table[D[ξ, χ[[i]]], {i, Length[χ]}]

F[{x_, y_, z_}] = grad[x y z - x - y - z, {x, y, z}]

{-1 + y z, -1 + x z, -1 + x y}
```

The unit sphere is parametrized with spherical coordinates by

```
r[φ_, θ_] := {Cos[θ] Sin[φ], Sin[θ] Sin[φ], Cos[φ]}
```

and the surface S corresponds to $0 \le \phi \le \pi/2$ and $0 \le \theta \le \pi/2$. The function of ϕ and θ that we need to integrate is

```
ξ[φ_, θ_] = F[r[φ, θ]].(∂φ r[φ, θ] × ∂θ r[φ, θ]) // Simplify
```

$\text{Cos}[\phi]^2 \, \text{Sin}[\theta] - 2\,\text{Cos}[\phi]\,\text{Sin}[\phi] - \text{Sin}[\theta]\,\text{Sin}[\phi]^2 + \text{Cos}[\theta]$
$\left(\text{Cos}[\phi]^2 - 3\,\text{Cos}[\phi]^3\,\text{Sin}[\theta]\,\text{Sin}[\phi] - \text{Sin}[\phi]^2 + \text{Cos}[\phi]\,\text{Sin}[\theta]\,\text{Sin}[\phi]\,\left(1 + 3\,\text{Sin}[\phi]^2\right)\right)$

Thus, the integral of ∇f over S is

$$\int_0^{\pi/2} \int_0^{\pi/2} \xi[\theta, \phi] \, d\theta \, d\phi$$

$$\frac{1}{4}\,(1 - 2\,\pi)$$

This is a plot of the vector field and the surface:

```
Show[VectorFieldPlot3D[F[{x, y, z}],
    {x, .15, 1}, {y, .15, 1}, {z, .15, 1}, PlotPoints → 5,
    VectorHeads → True, ScaleFactor → .4],
   ParametricPlot3D[r[φ, θ], {φ, 0, π/2}, {θ, 0, π/2}]]
```

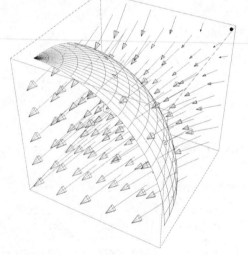

■ Example 6.4.6

Let **F** *be the vector field defined by*

$$F[\{x\_, y\_, z\_\}] := \frac{\{x + z, \, x + y, \, y + z\}}{\sqrt{x^2 + y^2 + z^2}}$$

and let S be the cylindrical surface given by $x^2 + y^2 = 1$ *and* $0 \le z \le 1$. *Compute the flux of* **F** *across S, and plot the vector field and the surface.*

A parametrization of the surface (using cylindrical coordinates) is

```
r[θ_, z_] := {Cos[θ], Sin[θ], z}
```

where $0 \le \theta \le 2\pi$ and $0 \le z \le 1$. So the integrand we need is

```
ξ[θ_, z_] = F[r[θ, z]].(∂_θ r[θ, z] × ∂_z r[θ, z]) // Simplify
```

$$\frac{1 + \text{Cos}[\theta] \ (z + \text{Sin}[\theta])}{\sqrt{1 + z^2}}$$

and the resulting integral is

```
{int = ∫₀¹ ∫₀²π ξ[θ, z] dθ dz, N[int]}
```

```
{2 π ArcSinh[1], 5.53783}
```

The following is a plot of the vector field and the surface:

```
Show[VectorFieldPlot3D[F[{x, y, z}],
  {x, -1, 1}, {y, -1, 1}, {z, 0, 1}, PlotPoints → 6,
  VectorHeads → True, ScaleFactor → .3],
ParametricPlot3D[r[θ, z], {θ, 0, 2π}, {z, 0, 1}, Mesh → None]]
```

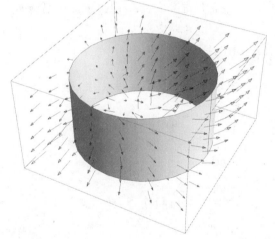

◆ Exercises

17. Find the area of the portion of the surface $z = x^2 + e^y$ that lies above:

    a) the disk $x^2 + y^2 \le 1$;   b) the square given by $-1 \le x \le 1$, $-1 \le y \le 1$.

18. A thin spherical shell, with radius 1, has mass density given by the square of the distance from one fixed point on the shell. Find the mass of the shell. (Hint: Think of the shell as the unit sphere and assume that the fixed point from which distance is measured is $(0, 0, -1)$.)

19. A thin metal sheet is bent in the shape of the portion of the surface $z = x^2 + y^2$ that lies above the square $-1 \le x \le 1, -1 \le y \le 1$. Its mass density is $\rho(x, y, z) = z$. Find its mass.

20. A thin metal sheet is bent in the shape of the portion of the surface $z = x^2 + y^2$ that lies above the disk $x^2 + y^2 \le 1$. Its mass density is $\rho(x, y, z) = z$. Find its mass.

21. Let $f(x, y, z) = xyz$. Find the flux of ∇f across the portion of the plane $2x + 3y + z = 6$ on which $x \ge 0$, $y \ge 0$, and $z \ge 0$.

22. Let $\mathbf{F}(x, y, z) = \langle -y, x, z \rangle$ and let S be the portion of the surface $z = x^2 y^2$ that lies above the unit quarter disk in the first quadrant of the xy-plane. Compute the flux of $\mathbf{F}$ across S.

23. Let $\mathbf{F}(x, y, z) = \langle x, y, z \rangle$ and S be the sphere $x^2 + y^2 + (z - 1)^2 = 1$. Compute the flux of $\mathbf{F}$ across S.

24. Let $\mathbf{F}(x, y, z) = \langle x, y, z \rangle$ and let S be the cylindrical surface given by $x^2 + y^2 = 1$ and $0 \le z \le 1$. Compute the flux of $\mathbf{F}$ across S.

6.5 Divergence, Curl, and the Laplacian

See Section 17.5 of Stewart's *Calculus* for definition and discussion of the divergence and curl of a vector field and the Laplacian of a function. (ET:16.5)

■ Divergence

The **divergence** of a two-dimensional vector field $\mathbf{F}(x, y) = \langle p(x, y), q(x, y) \rangle$ is the scalar function defined by

$$\text{div}\,\mathbf{F} = \nabla \cdot \mathbf{F} = \frac{\partial p}{\partial x} + \frac{\partial q}{\partial y}.$$

The divergence of a three-dimensional vector field $\mathbf{F}(x, y, z) = \langle p(x, y, z), q(x, y, z), r(x, y, z) \rangle$ is the scalar function defined by

$$\text{div}\,\mathbf{F} = \nabla \cdot \mathbf{F} = \frac{\partial p}{\partial x} + \frac{\partial q}{\partial y} + \frac{\partial r}{\partial z}.$$

The divergence of a vector field in any dimension is:

```
                        Length[x]
    div[F_, x_] :=         ∑       D[ F[[i]], x[[i]] ]
                          i=1
```

The first argument $\mathbf{F}$ is the expression representing the vector field, and the second argument x is the list of variables. Here's a three-dimensional example:

```
    div[{x y, y z, Cos[x y z]}, {x, y, z}]

    y + z - x y Sin[x y z]
```

■ Curl

The **curl** of a three-dimensional vector field $\mathbf{F}(x, y, z) = \langle p, q, r \rangle$ is the vector function defined by

$$\text{curl}\,\mathbf{F} = \nabla \times \mathbf{F} = \left\langle \frac{\partial r}{\partial y} - \frac{\partial q}{\partial z}, \frac{\partial p}{\partial z} - \frac{\partial r}{\partial x}, \frac{\partial q}{\partial x} - \frac{\partial p}{\partial y} \right\rangle.$$

A *Mathematica* definition of the curl of a three-dimensional vector field is:

```
curl[{p_, q_, r_}, {x_, y_, z_}] := {∂_y r - ∂_z q, ∂_z p - ∂_x r, ∂_x q - ∂_y p}
```

Here's an example of its usage:

```
curl[{x y, y z, x y z}, {x, y, z}]

{-y + x z, -y z, -x}
```

■ The Laplacian

The **Laplacian** of a function $f(x, y)$ is

$$\nabla^2 f(x, y) = \nabla \cdot \nabla f(x, y) = f_{xx}(x, y) + f_{yy}(x, y).$$

Similarly, the Laplacian of a function $f(x, y, z)$ is

$$\nabla^2 f(x, y, z) = \nabla \cdot \nabla f(x, y, z) = f_{xx}(x, y, z) + f_{yy}(x, y, z) + f_{zz}(x, y, z).$$

⚠ **Note:** While physicists are fond of using the notation $\nabla^2 f$ for the Laplacian of f, the more standard notation among mathematicians is Δf, since the notation $\nabla^2 f$ is also commonly used for the *Hessian matrix* of f, whose rows consist of the gradients of the components of ∇f.

Recall our `grad` function for computing gradients from Section 4.4:

```
grad[f_, x_] := Table[D[f, x[[i]]], {i, Length[x]}]
```

With this and `div`, we can make the following *Mathematica* definition of the Laplacian of a function of any number of variables.

```
laplacian[f_, x_] := div[grad[f, x], x]
```

Here's an example with a function of three variables:

```
laplacian[x² y³ Cos[z], {x, y, z}] // Simplify
```

$$y \left(2 y^2 - x^2 \left(-6 + y^2\right)\right) Cos[z]$$

6.6 Stokes' Theorem

Stokes' theorem is the subject of Section 17.8 of Stewart's *Calculus*. (ET:16.8)

Stokes' theorem is a three-dimensional extension of Green's theorem. While Green's theorem provides a connection between line integrals of vector fields and double integrals in the plane, Stokes' theorem provides the following connection between line integrals and surface integrals of vector fields in three-dimensional space:

$$\int_C \mathbf{F} \cdot d\mathbf{r} = \iint_S \text{curl}\, \mathbf{F} \cdot d\mathbf{S}.$$

Here S is assumed to be an oriented, piecewise-smooth surface whose boundary (or "edge") is a simple, closed, piecewise-smooth, positively oriented curve C. The vector field $\mathbf{F}$ is assumed to be continuously differentiable on an open region containing S.

- ## Example 6.6.1

Verify the conclusion of Stokes' theorem if S is the portion of the paraboloid $z = x^2 + y^2$ that lies under the plane $z = 1$ and **F** *is the vector field*

$$\mathbf{F}[\{x\_, y\_, z\_\}] := \{y - z^2, z - x^2, x - y^2\}$$

Let's first parametrize S by

$$\mathbf{p}[r\_, \theta\_] := \{r \, \text{Cos}[\theta], r \, \text{Sin}[\theta], r^2\}$$

With this parametrization, the normal vector

$$\mathbf{n}[r\_, \theta\_] = (\partial_r \mathbf{p}[r, \theta]) \times (\partial_\theta \mathbf{p}[r, \theta]) \; // \; \text{Simplify}$$

$$\{-2 \, r^2 \, \text{Cos}[\theta], -2 \, r^2 \, \text{Sin}[\theta], r\}$$

points upward and toward the z-axis. The curve C of interest here is the unit circle in the plane, lifted to a height $z = 1$. Since C should have a positive orientation relative to S, we need to use a counterclockwise orientation of the unit circle in our parametrization. So we'll parametrize C by

$$\gamma[t\_] := \{\text{Cos}[t], \text{Sin}[t], 1\}$$

The line integral of **F** along C is then

$$\int_0^{2\pi} \mathbf{F}[\gamma[t]].\gamma'[t] \; dt$$

$-\pi$

The curl of **F** is

$$\text{curlF}[\{x\_, y\_, z\_\}] = \text{curl}[\mathbf{F}[\{x, y, z\}], \{x, y, z\}]$$

$$\{-1 - 2 \, y, -1 - 2 \, z, -1 - 2 \, x\}$$

and so the surface integral of interest is

$$\int_0^{2\pi} \int_0^1 \text{curlF}[\mathbf{p}[r, \theta]].\mathbf{n}[r, \theta] \; dr \, d\theta$$

$-\pi$

which is the same as the line integral of **F** along C.

- ## Example 6.6.2

Use Stokes' theorem to evaluate the line integral of

$$\mathbf{F}[\{x\_, y\_, z\_\}] := \{y \, z + \text{Cos}[\pi x], z - \text{Sin}[\pi y], x \, y + z^3\}$$

along C, *where* C *is the intersection of the parabolic sheet* $z = x^2$ *and the cylinder* $x^2 + y^2 = 1$, *oriented in a way that corresponds to a counterclockwise orientation of its projection onto the xy-plane.*

Rather than parametrize C and compute the line integral of $\mathbf{F}$ along C, which would not be easy to do, we will appeal to Stokes' theorem and compute instead the surface integral of curl $\mathbf{F}$ over the portion of the surface $z = x^2$ on which $x^2 + y^2 \leq 1$. The curl of $\mathbf{F}$ is

```
curlF[{x_, y_, z_}] = curl[F[{x, y, z}], {x, y, z}]
```

```
{-1 + x, 0, -z}
```

A parametrization of the surface $z = x^2$ is

```
q[x_, y_] := {x, y, x²}
```

The following is a plot of the two surfaces that indicates the curve C:

```
Show[ParametricPlot3D[q[x, y], {x, -1.1, 1.1}, {y, -1.3, 1.3}],
    ParametricPlot3D[{Cos[θ], Sin[θ], z}, {θ, 0, 2 π}, {z, -.2, 1.2}]]
```

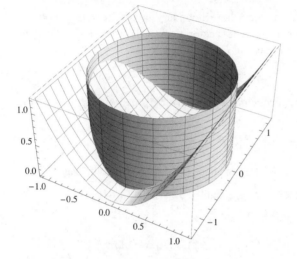

A normal vector to the surface $z = x^2$ is

```
n[x_, y_] = (∂ₓq[x, y]) × (∂ᵧq[x, y])
```

```
{-2 x, 0, 1}
```

Since this normal vector points upward, we know we have the proper orientation. So we need to integrate the function

```
ξ[x_, y_] = curlF[q[x, y]].n[x, y]
```

$$-2 (-1 + x) x - x^2$$

over the unit disk in the xy-plane. Converting to polar coordinates, we find

```
∫₀²π ∫₀¹ ξ[r Cos[θ], r Sin[θ]] r dr dθ
```

$$-\frac{3\pi}{4}$$

6.7 The Divergence Theorem

Section 17.9 of Stewart's *Calculus* states—and proves a special case of—the divergence theorem. (ET:16.9)

Another extension of Green's theorem to three-dimensions is the divergence theorem (a.k.a. Gauss's theorem). While Green's theorem provides a connection between line integrals of vector fields and double integrals of scalar functions in the plane, the divergence theorem provides the following connection between surface integrals of vector fields and triple integrals of scalar functions in three-dimensional space:

$$\iint_{\partial E} \mathbf{F} \cdot d\mathbf{S} = \iint_E \operatorname{div} \mathbf{F} \, dV.$$

Here E is assumed to be a solid region whose boundary is a piecewise-smooth surface ∂E with a positive (outward) orientation. The vector field $\mathbf{F}$ is assumed to be continuously differentiable on an open region containing E.

■ **Example 6.7.1**

Verify the conclusion of the divergence theorem where E is the unit ball and $\mathbf{F}$ is the vector field

```
F[{x_, y_, z_}] := {y - x, x² + y² + z², (y - z)²}
```

Let's first parametrize ∂E, which is the unit sphere, with spherical coordinates:

```
p[φ_, θ_] := {Cos[θ] Sin[φ], Sin[θ] Sin[φ], Cos[φ]}
```

and compute the *outward* normal vector:

```
n[φ_, θ_] = (∂φ p[φ, θ]) × (∂θ p[φ, θ]) // Simplify
```

$$\{Cos[\theta] Sin[\phi]^2, Sin[\theta] Sin[\phi]^2, Cos[\phi] Sin[\phi]\}$$

Thus, $\mathbf{F}$ restricted to ∂E is

```
ξ[φ_, θ_] = F[p[φ, θ]] // Simplify
```

$$\{(-Cos[\theta] + Sin[\theta]) Sin[\phi], 1, (Cos[\phi] - Sin[\theta] Sin[\phi])^2\}$$

So the flux of $\mathbf{F}$ across ∂E is

```
∫₀²π ∫₀π ξ[φ, θ].n[φ, θ] dφ dθ
```

$$-\frac{4\pi}{3}$$

The divergence of $\mathbf{F}$ is

```
divF[x_, y_, z_] = div[F[{x, y, z}], {x, y, z}] // Simplify
```

$$-1 + 2z$$

which in spherical coordinates becomes

$$\xi[\rho\_, \phi\_, \theta\_] = \text{divF}[\rho \, \text{Cos}[\theta] \, \text{Sin}[\phi], \, \rho \, \text{Sin}[\theta] \, \text{Sin}[\phi], \, \rho \, \text{Cos}[\phi]]$$

$$-1 + 2 \, \rho \, \text{Cos}[\phi]$$

So the integral of div **F** over the unit sphere is

$$\int_0^{2\pi} \int_0^{\pi} \int_0^1 \xi[\rho, \phi, \theta] \, \rho^2 \, \text{Sin}[\phi] \, d\rho \, d\phi \, d\theta$$

$$-\frac{4\pi}{3}$$

which is the same as the integral of the flux of **F** across ∂E.

■ Example 6.7.2

Use the divergence theorem to compute the flux of

$$\text{F}[\{x\_, y\_, z\_\}] := \left\{ x^2 + \text{Cos}\left[y^2 \, z\right], \, x - y \, z + z, \, x \, y \, z \right\}$$

across the surface of the box E described by $0 \le x \le 1, \, 0 \le y \le 1, \, 0 \le z \le 1.$

The divergence of **F** is

$$\text{divF}[x\_, y\_, z\_] = \text{div}[\text{F}[\{x, y, z\}], \{x, y, z\}]$$

$$2 \, x + x \, y - z$$

By the divergence theorem, the flux across E (with outward normal vector) is the same as the triple integral of div **F** over E :

$$\int_0^1 \int_0^1 \int_0^1 \text{divF}[x, y, z] \, dx \, dy \, dz$$

$$\frac{3}{4}$$

◆ Exercises

25. Use Stokes' theorem to evaluate the line integral of

$$\text{F}(x, y, z) = \left\langle y \, z - \cos \pi x, \, z - \sin \pi y, \, x \, y + z^3 \right\rangle$$

along the boundary of the parallelogram that lies in the plane $x + y + z = 2$ and above the unit square in the xy-plane, oriented so that it corresponds to a counterclockwise orientation of the edges of the unit square in the xy-plane.

26. Use Stokes' theorem to evaluate the line integral of

$$\text{F}(x, y, z) = \left\langle z - \cos x, \, x^2 - z^2, \, x^2 + y^2 + z^2 \right\rangle$$

along C, where C is the intersection of the surface $z = x^2 + y^3$ and the cylinder $x^2 + y^2 = 1$, oriented in a way that corresponds to a counterclockwise orientation of its projection onto the xy-plane.

27. Use Stokes' theorem to evaluate the line integral of

$$\mathbf{F}(x, y, z) = \left\langle z + x^5, \ x + y^{2/3}, \ xz + \cos(\pi y) \right\rangle$$

along C, where C is the intersection of the surface $z = 1 - x^2 y$ and the surface of the box given by $-1 \le x \le 1$, $-1 \le y \le 1$, $0 \le z \le 2$, oriented in a way that corresponds to a counterclockwise orientation of its projection onto the xy-plane.

28. Verify the conclusion of the divergence theorem where

$$\mathbf{F}(x, y, z) = \left\langle xy, \ y^2, \ z + xy \right\rangle$$

and E is the cylinder given by $x^2 + y^2 \le 1$ and $0 \le z \le 1$. (Note: The surface ∂E has three parts.)

29. Use the divergence theorem to compute the flux of

$$\mathbf{F}(x, y, z) = \left\langle x^2 - yz, \ y^2 + xz, \ z^2 - xy \right\rangle$$

across the surface of the cylinder given by $x^2 + y^2 \le 1$ and $0 \le z \le 1$.

30. Let E be the hemisphere described by $x^2 + y^2 + z^2 \le 1$ and $z \ge 0$. The flux of

$$\mathbf{F}(x, y, z) = \left\langle x^2 z + \cos(\pi z), \ y - 2xyz, \ -z + x + y \right\rangle$$

across ∂E is zero according to the divergence theorem, since div $\mathbf{F} = 0$. Make use of this to help compute the flux of $\mathbf{F}$ across the spherical (top) part of the surface ∂E.

31. Suppose that a function f satisfies $\nabla^2 f(x, y, z) = e^{-(x^2+y^2+z^2)}$ in the interior of the unit sphere. Find the flux of ∇f across the surface of the unit sphere.

7 Projects

7.1 Osculating Circles

> Background material is found in Sections 14.1-14.3 of Stewart's *Calculus*. (ET:13.1-3)

Let C be a curve in the plane, parametrized by $\gamma(t) = \langle x(t), y(t) \rangle$. Recall from Section 3.4 our *Mathematica* definitions of the unit tangent vector $\mathbf{T}(t)$:

$$\texttt{unitTan[}\gamma\texttt{\_Symbol, t\_Symbol] := } \gamma\texttt{'[t]} \Big/ \sqrt{\gamma\texttt{'[t]}.\gamma\texttt{'[t]}}$$

and curvature $\kappa(t)$:

$$\texttt{kappa[}\gamma\texttt{\_, t\_Symbol] := } \sqrt{\frac{\texttt{D[unitTan[}\gamma\texttt{, t], t].D[unitTan[}\gamma\texttt{, t], t]}}{\gamma\texttt{'[t]}.\gamma\texttt{'[t]}}}$$

One interpretation of curvature is that its reciprocal is the radius of the circle that best approximates the curve at the given point. However, since $\kappa(t)$ is by definition a nonnegative quantity, it does not indicate on which side of the curve this best approximating circle should lie—i.e., it only indicates the *amount* of "bend" in the curve and does not indicate which *way* the curve is bending. So it will be useful to define a **signed curvature** $\kappa_\pm(t)$ by

$$\texttt{sgndKappa[}\gamma\texttt{\_, t\_Symbol] := kappa[}\gamma\texttt{, t] Sign}\big[\texttt{Det}\big[\{\gamma\texttt{'[t], }\gamma\texttt{''[t]}\}\big]\big]$$

---

1. Explain carefully why the signed curvature will be positive when the curve bends to the left and negative when it bends to the right. (*Hint:* Think about the cross product of $\langle x'(t), y'(t), 0 \rangle$ and $\langle x''(t), y''(t), 0 \rangle$.)

The **osculating circle** to the curve at a point $\gamma(t_0)$ is the circle that best approximates the curve near that point. At the given point, it is tangent to the curve, has the same curvature, and lies on the side of the curve toward which the curve is bending. The following plot shows osculating circles to the curve $y = \tan^{-1}(3x)$ at $x = \pm .5$.

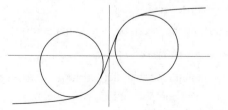

2. Show that the position vector of the center of the osculating circle at $\gamma(t)$ is

$$\langle x_0, y_0 \rangle = \gamma(t) + \frac{1}{\kappa_{\pm}(t)} \mathbf{n}(t),$$

where $\mathbf{n}(t)$ is the left-pointing unit normal vector to the curve given by

$$\mathbf{n}(t) = \frac{\langle -y'(t), x'(t) \rangle}{\sqrt{x'(t)^2 + y'(t)^2}}.$$

3. Create a plot that shows the curve $\gamma(t) = \langle t, \sin t \rangle$ along with osculating circles at $(\pi/2, 1)$ and $(3\pi/2, -1)$.

4. Create a *Mathematica* "program" `oscircle[{x_,y_},{t_,t1_,t2_},t0_]` that will plot the curve parametrized by $x = x(t)$, $y = y(t)$ for $t_1 \le t \le t_2$ along with the osculating circle at $(x(t_0), y(t_0))$. The program should be a `Module` containing the computation of the center of the circle and the `ParametricPlot` command that plots the circle. Use it to plot osculating circles on the curve $\gamma(t) = \langle \sin t, \sin 2t \rangle$ at points corresponding to $t = \pi/5, \pi/4,$ and $\pi/3$.

5. Create an animation of the osculating circle rolling along the curve

$$\gamma(t) = \frac{10}{10 + t^2} \langle \cos t, \sin t \rangle, \ 0 \le t \le 2\pi.$$

7.2 Coriolis Acceleration

Background material is found in Section 14.4 of Stewart's *Calculus*. (ET:13.4)

Consider a disk of radius 1, centered at the origin, rotating about its center with constant angular speed ω. The motion of the point P_0 that is located at $(1, 0)$ at time $t = 0$ could be parametrized by $\mathbf{p}_0(t) = \langle \cos \omega t, \sin \omega t \rangle$. Suppose that a bug begins moving at time $t = 0$ at constant speed c from the origin toward the point P_0 on the edge of the disk. The bug thinks it is moving in a straight line on the disk, but to a stationary observer overhead, the bug is moving along the curved path parametrized by

$$\gamma(t) = c t \, \mathbf{p}_0(t) \text{ for } 0 \le t \le 1/c.$$

1. Let $\omega = 2\pi$. Plot the path of the bug and the path of the destination point P_0 for each of $c = 1/2, 1, 2, 4,$ and 8. Include a few of the bug's velocity vectors in each plot. (Use `Vector` from the `Vectors` package or `Arrow`.)

2. Compute the bug's acceleration and show that it can be written in the form

$$\mathbf{a}(t) = 2 \, \mathbf{p}_0'(t) - \omega^2 t \, \mathbf{p}_0(t).$$

Note that this consists of a component that is parallel to the (radial) vector $\mathbf{p}_0(t)$ and a component that is orthogonal to $\mathbf{p}_0(t)$. The component that is parallel to $\mathbf{p}_0(t)$ is the **centripetal acceleration** and the component that is orthogonal to $\mathbf{p}_0(t)$ is the **Coriolis acceleration**.

Redo the plots in #1 without the velocity vectors, but showing the centripetal and Coriolis acceleration vectors at several points along the bug's path.

Consider now a sphere of radius 1 centered at the origin, that rotates about the z-axis in a "west-to-east" direction with constant angular speed ω. Suppose that a bug travels due "north" on the sphere with angular speed η and is located at the point $(1, 0, 0)$ on the "equator" at time $t = 0$. Then the position of the bug is given by

$$\gamma(t) = \langle \cos \eta t \cos \omega t, \cos \eta t \sin \omega t, \sin \eta t \rangle \text{ for } -\frac{\pi}{2\eta} \le t \le \frac{\pi}{2\eta}.$$

3. Compute the acceleration and show that it can be written as

$$\mathbf{a}(t) = \mathbf{a}_0(t) + \mathbf{a}_c(t) + \mathbf{a_k}(t),$$

where

$$\mathbf{a}_0(t) = -\left(\omega^2 + \eta^2\right) \cos(\eta t) \langle \cos(\omega t), \sin(\omega t), 0 \rangle;$$

$$\mathbf{a}_c(t) = -2 \omega \eta \sin(\eta t) \langle -\sin(\omega t), \cos(\omega t), 0 \rangle;$$

$$\mathbf{a_k}(t) = \langle 0, 0, -\eta^2 \sin(\eta t) \rangle.$$

4. Show that:

a) $\mathbf{a}_0(t)$ is orthogonal to and points toward the z-axis;

b) $\mathbf{a_k}(t)$ is parallel to the z-axis, pointing upward in the southern hemisphere and downward in the northern hemisphere;

c) $\mathbf{a}_c(t)$ is orthogonal to $\mathbf{a}_0(t)$, $\mathbf{a_k}(t)$, and $\mathbf{r}(t)$. Thus, $\mathbf{a}_c(t)$ is orthogonal to the sphere and horizontal—i.e., parallel to the xy-plane. This is the Coriolis acceleration vector in three dimensions.

5. Create a "wire frame" image of the unit sphere as follows:

```
wireSphere = SphericalPlot3D[1, {ϕ, 0, π},
    {θ, 0, 2 π}, Mesh → {12, 24}, PlotStyle → None];
Show[wireSphere, Boxed → False, Axes → None]
```

Then (using **Vector** from the **Vectors** package) create a plot that shows **wireSphere** plus:

a) the curve $\gamma(t) = \langle \cos(\pi t/2) \cos \pi t, \cos(\pi t/2) \sin \pi t, \sin(\pi t/2) \rangle$, $-1 \le t \le 1$;

b) the vectors $\mathbf{a}_0(.5)$, $\mathbf{a}_c(.5)$, and $\mathbf{a_k}(.5)$;

c) the vectors $\mathbf{a}_0(-.5)$, $\mathbf{a}_c(-.5)$, and $\mathbf{a_k}(-.5)$.

Suggestion: Scale the $\mathbf{a}_0$ vectors by a factor of 0.1 and the $\mathbf{a}_c$ vectors by 0.25.

6. In your plot from #5, observe the direction (east or west) of the Coriolis acceleration in each of the southern and northern hemispheres. Show that this observation would be the same if the bug travels north-to-south, rather than south-to-north.

7.3 An Ant on a Helix

> Background material is found in Sections 14.1-14.3 of Stewart's *Calculus*. (ET:14.1-3)

Suppose that an ant walks along a helical path on the surface of a transparent cylinder $x^2 + y^2 = 1$ such that its position vector at time t is $\mathbf{r}(t) = \langle \cos t, \sin t, t \rangle$. A light source is located on the z-axis at the point $(0, 0, 8\pi)$, causing the ant to cast a shadow on the xy-plane while $0 \le t \le 8\pi$.

1. Find a parametrization of the curve in the xy-plane traced by the ant's shadow and plot its graph. Create a second, three-dimensional plot showing the helix and the path of the shadow on the xy-plane.

2. Find the length of the curve traced by the ant's shadow while $0 \le t \le 4\pi$.

7.4 Animating Particle Motion

> Background material is found in Chapter 14 of Stewart's *Calculus*. (ET:13)

1. To begin with a simple example, enter the following surface animation. Then adjust the speed with the buttons at the top of the panel.

$$\texttt{Animate}\left[\texttt{Plot3D}\left[\texttt{Cos}\left[\sqrt{\texttt{x}^2 + \texttt{y}^2} + \texttt{k}\right], \{\texttt{x, -3, 3}\},\right.\right.$$

$$\left.\left.\{\texttt{y, -3, 3}\}, \texttt{Axes} \rightarrow \texttt{False, Mesh} \rightarrow \texttt{True}\right], \{\texttt{k, 0, 2}\pi\}\right]$$

The following defines a command named `motion`, which creates an animation of a curve being traced out by a moving particle while showing the velocity vector at each position of the particle.

```
Needs["Vectors"];
motion[γ_,{t_,t1_,t2_},{x_,a_,b_},{y_,c_,d_},opts___Rule] :=
  Animate[Show[ParametricPlot[γ, {t, t1, k + .0001},
    PlotStyle → {Thick, Red}], Vector[{γ, D[γ, t]} /. t → k],
    Graphics[{PointSize[Large], Point[γ /. t → k]}], opts,
    Axes → False, PlotRange → {{a, b}, {c, d}}], {k, t1, t2}]
```

In the statement

```
motion[γ, {t, t1, t2}, {x, a, b}, {y, c, d}]
```

$γ$ (gamma) is an expression that gives the parameterization of the curve in terms of the variable t, which ranges from $t1$ to $t2$. Each plot is drawn over the rectangle described by $a \le x \le b$, $c \le y \le d$.

---

2. To illustrate, let's use the following curve:

```
r[t_] := {2 Cos[t] + Cos[5 t], 2 Sin[t] + Sin[5 t]};
ParametricPlot[r[t], {t, 0, 2 π}, Axes → False]
```

The following command will produce the animation. Enter it, and then adjust the speed of the animation as you like.

```
motion[r[t], {t, 0, 2 π}, {x, -7, 7}, {y, -7, 7.2}]
```

3. Create animations similar to the one in #2 for each of the following parametrizations of the top half of the unit circle.

 a) $\mathbf{r}(t) = \langle \cos t, \sin t \rangle,\ 0 \le t \le \pi$

 b) $\mathbf{r}(t) = \langle \cos 2t, \sin 2t \rangle,\ 0 \le t \le \pi/2$

 c) $\mathbf{r}(t) = \langle \cos t^2, \sin t^2 \rangle,\ 0 \le t \le \sqrt{\pi}$

4. Create animations similar to the one in #2 for each of the following curves.

 a) $\mathbf{r}(t) = \langle \sin t, \sin 2t \rangle,\ 0 \le t \le \pi$

 b) $\mathbf{r}(t) = \langle t, t - t^2 \rangle,\ 0 \le t \le 1$

 c) $\mathbf{r}(t) = \langle t \cos t, t \sin t \rangle,\ 0 \le t \le 2\pi$

5. Modify `motion` so that it plots acceleration vectors in addition to velocity vectors. Then redo parts 2-4.

7.5 Flying the Osculating Plane

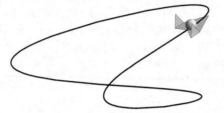

Background material is found in Chapter 14 of Stewart's *Calculus*. (ET:13)

Suppose that a particle moves along a curve in three dimensions and has the position vector $\mathbf{r}(t) = \langle x(t), y(t), z(t) \rangle$ at time t. The velocity and acceleration vectors are, respectively,

$$\mathtt{v[t\_] := r'[t]; \; a[t\_] := v'[t]}$$

The unit tangent vector $\mathbf{T}(t)$ in the direction of the velocity is given by

$$\mathtt{T1[t\_] := \frac{v[t]}{\sqrt{v[t].v[t]}}}$$

and the curvature of the path is given by

$$\kappa(t) = \frac{|\mathbf{T}'(t)|}{|\mathbf{v}(t)|}.$$

The speed $v(t)$ of the particle is $v(t) = |\mathbf{v}(t)|$. For this project we will also need a unit normal vector to the curve that is given by

$$\mathbf{N}(t) = \frac{\mathbf{T}'(t)}{|\mathbf{T}'(t)|},$$

for which we will give the following *Mathematica* definition:

$$\mathtt{N1[t\_] := Module\left[\{dTdt = D[unitTan[r, t], t]\}, \frac{dTdt}{\sqrt{dTdt.dTdt}}\right]}$$

It is shown in Section 14.4 of Stewart's *Calculus* that the acceleration vector $\mathbf{a}(t) = \mathbf{v}'(t)$ can be written as

$$\mathbf{a}(t) = v(t)\,\mathbf{T}(t) + \kappa(t)\,v(t)^2\,\mathbf{N}(t),$$

which shows that

> $\mathbf{a}(t)$ *always lies in the plane that is parallel to* $\mathbf{T}(t)$ *and* $\mathbf{N}(t)$.

The plane parallel to $\mathbf{T}(t)$ and $\mathbf{N}(t)$ and containing the point with position vector $\mathbf{r}(t)$ is call the **osculating plane** for the curve at $\mathbf{r}(t)$. The osculating plane is always tangent to the curve and parallel to the acceleration of the particle.

Plot the path of the particle whose position vector at time t is

```
r[t_] := {Sin[t] + Cos[2 t], Cos[2 t], Cos[t]}
```

2. Replot the path in #1 with `Axes→False` and `Boxed→False`, naming the plot "path." Plot the curve in color if you like.

3. Check that `T1[t].N1[t]=0` for all t.

4. A small "patch" of the osculating plane at the point with position vector $\mathbf{r}(t)$ is given by

```
plane[t_] := Graphics3D[Polygon[{r[t] + .3 N1[t] - .1 T1[t],
    r[t], r[t] - .3 N1[t] - .1 T1[t], r[t] + .5 T1[t]}]]
```

Plot `plane[π/6]` along with `path`.

5. Imagine that the triangular patch is your airplane. Create an animation of the flight of your airplane as t goes from 0 to 2π. Adjust the speed of the animation to your liking. (Suggestion: Use `PlotRange→{{-2.3,1.7},{-1.2,1.4},{-1.2,1.3}}` for this curve.

6. Fly your plane along the following curves. Adjust the length and wingspan of your plane as you like.

   a) $\mathbf{r}(t) = \langle \cos 2t, 2 \sin 2t, \sin t \rangle$ b) $\mathbf{r}(t) = \langle \cos t, \sin 2t, \cos(t + 2\pi/3) \rangle$

   c) $\mathbf{r}(t) = \langle \cos t, \sin t, \cos t \sin t \rangle$ d) $\mathbf{r}(t) = \langle \cos t, \sin t, \cos t \sin 2t \rangle$

❀ **Bonus**: Redo the above after (a) using `Sphere` to add a "cockpit" to your airplane and (b) giving your airplane turned-up "*Gulfstream*" wingtips as in the picture at the beginning of this project. (Use the *binormal vector* $\mathbf{B}(t) = \mathbf{T}(t) \times \mathbf{N}(t)$.)

7.6 Drawing Curves on a Sphere

Background material is found in Sections 14.1 and 14.4 of Stewart's *Calculus*. (ET:13.1, 13.4)

Recall that the usual parametrization of the unit sphere is

$$\mathbf{r}(\theta, \phi) = \langle \cos\theta \sin\phi, \ \sin\theta \sin\phi, \ \cos\phi \rangle.$$

There are various ways to make use of this to plot space curves that lie on the unit sphere. The most obvious way is simply to substitute functions $\theta(t)$ and $\phi(t)$ into $\mathbf{r}(\theta, \phi)$, obtaining the curve

$$\gamma(t) = \langle \cos\theta(t) \sin\phi(t), \ \sin\theta(t) \sin\phi(t), \ \cos\phi(t) \rangle,$$

which is guaranteed to lie on the unit sphere.

Another way of generating interesting curves on the unit sphere is with a parametrization of the form

$$\gamma(t) = \left\langle \cos\theta(t) \sqrt{1 - z(t)^2}, \ \sin\theta(t) \sqrt{1 - z(t)^2}, \ z(t) \right\rangle$$

where $\theta(t)$ and $z(t)$ are chosen functions such that the values of $z(t)$ are between -1 and 1.

In the exercises below, you will be asked to plot several curves on the unit sphere. You may need to increase `PlotPoints` in order to get good plots, and you may also want to increase the thickness of the curve and/or plot it in color.

---

1. Show that, for a curve parametrized in either of the two ways described above, every point on the curve is on unit sphere.

2. Create a "wire frame" image of the unit sphere as follows:

   ```
   wireSphere = SphericalPlot3D[1, {ϕ, 0, π}, {θ, 0, 2π},
       Mesh → {12, 24}, MeshStyle → Gray, PlotStyle → None];
   Show[wireSphere, Boxed → False, Axes → None]
   ```

   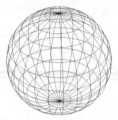

3. Plot each of the following curves and show the result together with `wireSphere` from #2.

   a) $\gamma(t) = \langle \cos t \sin t, \sin t \sin t, \cos t \rangle, \ 0 \le t \le 2\pi,$

   b) $\gamma(t) = \langle \cos t \sin 2t, \sin t \sin 2t, \cos 2t \rangle, \ 0 \le t \le 2\pi$

   c) $\gamma(t) = \langle \cos 6t \sin t, \sin 6t \sin t, \cos t \rangle, \ 0 \le t \le \pi$

4. Plot the curve parametrized by

   $$\gamma(t) = \left\langle \cos(\theta(t)) \sqrt{1 - z(t)^2}, \ \sin(\theta(t)) \sqrt{1 - z(t)^2}, \ z(t) \right\rangle$$

   for each of the following cases. Show each result along with `wireSphere` from #2.

   a) $\theta(t) = -5\pi t, \ z(t) = \sqrt{1 - t^4}, \ 0 \le t \le 1$

   b) $\theta(t) = -5\pi t, \ z(t) = 1/(t + 1), \ 0 \le t \le 1$

   c) $\theta(t) = 4\pi t, \ z(t) = \dfrac{2}{\pi} \tan^{-1} t, \ -1 \le t \le 1$

5. Try to reproduce the following plots. The main difference between them is that in the plot on the left, $z(t)$ is a linear function of t.

   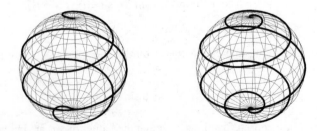

6. The following is a wire frame image of the cone $z = 2\left(x^2 + y^2\right)$.

```
wireCone = RevolutionPlot3D[2 r, {r, -1, 1}, {θ, 0, 2 π},
    MeshStyle → Gray, Mesh → {12, 24}, PlotStyle → None];
Show[wireCone, BoxRatios → Automatic, Boxed → False, Axes → None]
```

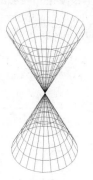

Try to reproduce the curve on the left below. Then experiment with various curves on this cone. Try to come up with something like the curve on the right below.

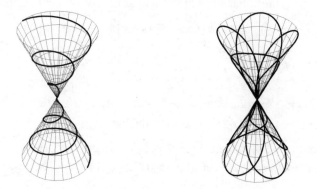

7.7 Steepest Descent Curves

Background material is found in Section 15.6 of Stewart's *Calculus*. (ET:14.6)

In this project, you will use *Mathematica* to plot families of steepest descent curves. An nice example to work with is provided by the function

$$f[x\_, y\_] := \left(x - y^2\right)^2 + (1 - y)^2$$

1. a) Plot the graph of f on the rectangle $-2 \le x \le 6$, $-2 \le y \le 2$, using `Plot3D` with the option `BoxRatios→{1,1,1}`.

   b) Create a contour plot of f on the rectangle $-2 \le x \le 6$, $-2 \le y \le 2$, with contours corresponding to $z = 0.02, 0.2, 1, 3, 5, 10, 20$. Use the option `ContourShading→ False` and name the plot "levels."

A parametric description $(x(t), y(t))$ of the steepest descent curves will satisfy the differential equations

$$x'(t) = -\partial_x f(x(t), y(t)) \text{ and } y'(t) = -\partial_y f(x(t), y(t)).$$

The ability of *Mathematica* to solve differential equations will help us plot these curves. To assemble the differential equations that we want to solve, we'll first compute the gradient of f, substituting $x(t)$ and $y(t)$ for x and y, respectively.

```
gradf =
  ({∂x f[x, y], ∂y f[x, y]} // Simplify) /. {x → x[t], y → y[t]}
```

$$\{2 \left(x[t] - y[t]^2\right), 2 \left(-1 + y[t] - 2 x[t] y[t] + 2 y[t]^3\right)\}$$

The differential equations are

```
diffeqs = Thread[{x'[t], y'[t]} == -gradf]
```

$$\{x'[t] == -2 \left(x[t] - y[t]^2\right), y'[t] == -2 \left(-1 + y[t] - 2 x[t] y[t] + 2 y[t]^3\right)\}$$

We also need to specify initial values for each of $x(t)$ and $y(t)$ as part of the system to be solved. This defines the initial-value problem as a function of the initial point (x_0, y_0):

```
ivp[{x0_, y0_}] = Join[diffeqs, {x[0] == x0, y[0] == y0}]
```

$$\{x'[t] == -2 \left(x[t] - y[t]^2\right),$$
$$y'[t] == -2 \left(-1 + y[t] - 2 x[t] y[t] + 2 y[t]^3\right), x[0] == x0, y[0] == y0\}$$

Next, let's define the solution of the initial-value problem over the time interval $(0, t_{end})$ as a function of the initial point (x_0, y_0) and the end time:

```
soln[{x0_, y0_}, endt_] := {x[t], y[t]} /. First[
    NDSolve[ivp[{x0, y0}], {x[t], y[t]}, {t, 0, endt}] ]
```

With all that in place, we can now plot, for instance, the solution curve starting at $(5, -1)$ over the time interval $(0, 4)$ as follows:

```
ParametricPlot[Evaluate[soln[{5, -1}, 4]],
  {t, 0, 6}, AspectRatio → Automatic]
```

We can plot at once several such curves, with initial points $(x_0, -3)$, $x_0 = -2, -1, \ldots, 5$, by entering the following:

```
endTime = 4; paths1 =
  Show[Table[ParametricPlot[Evaluate[soln[{x0, -2}, endTime]],
     {t, 0, endTime}, PlotStyle → {{Thickness[.006], Red}}],
     {x0, -2, 5}], PlotRange → All, AspectRatio → Automatic]
```

2. Show the above `paths1` plot atop the `levels` contour plot from #1.

3. (a) Create a similar plot named `paths2` by plotting curves with initial points $(x_0, 2)$, for $x_0 = -5, -2, ..., 5$. (b) Create plots called `paths3` and `paths4` by plotting curves with initial points (x_0, y_0), $x_0 = 2, ..., 5$, for each of $y_0 = -0.2$ and 0. (c) Show `paths1` through `paths4` along with `levels`.

4. Create a similar plot on the rectangle $-3 \le x \le 3$, $-2 \le y \le 2$, for

$$f(x, y) = 3 e^{-(x^2+y^2)} + \frac{1}{3}(x^2 + 2y^2).$$

Plot two collections of curves with initial points (x_0, y_0) for $y = \pm 2$ with $x = -3, -2.5, ...,$ 2.5, 3. Also plot the surface $z = f(x, y)$ over the same rectangle.

5. Create a similar plot on the square $-2 \le x \le 2$, $-2 \le y \le 2$, for each of

a) $f(x, y) = \frac{1}{2}(x^2 + y^2)$ b) $f(x, y) = \frac{1}{3}(x^2 + 2y^2)$ c) $f(x, y) = \frac{1}{4}(x^2 + 3y^2)$

In each part, two collections of curves with $y = \pm 2$ and $x = -2, -1, 0, 1, 2$ will suffice. However, this time extend the definition of `soln` to use `DSolve` rather than `NDSolve` by entering

```
soln[{x0_, y0_}] := {x[t], y[t]} /. First[
    DSolve[ivp[{x0, y0}], {x[t], y[t]}, t] ]
```

and remove the second argument (`endTime`) to `soln` inside `ParametricPlot`.

6. For each part of #5, enter

```
soln[{1, 1}]
```

and eliminate the parameter t in the result to obtain y as a function of x. What kind of curves are the steepest descent curves in each case? What general statement do you think can be made about the steepest descent curves for a quadratic function $f(x, y) = a(x^2 + k y^2)$?

7.8 Critical Point Classification

Background material is found in Section 15.7 of Stewart's *Calculus*. (ET:14.7)

■ Quadratic Functions

Any quadratic function f of n variables can be expressed in "vector form" as

$$f(\chi) = \tfrac{1}{2}\, \chi \cdot H\, \chi + \mathbf{a} \cdot \chi + b$$

where χ is the row vector containing the variables, $\mathbf{a}$ is a vector of length n, b is a number, and H is the $n \times n$ **Hessian matrix** of f, whose ij^{th} entry is

$$[H]_{i,j} = \frac{\partial^2 f}{\partial x_i\, \partial x_j}\,.$$

(The notation here blurs the distinction between row vectors and column vectors and so is somewhat contrary to traditional notation, but it is essentially consistent with *Mathematica* syntax, since the "Dot" operator is used for both matrix multiplication and the dot product of vectors.)

For example, with

```
H := {{2, 3}, {5, 6}}; χ := {x, y}; a := {7, -5}; b := 5;
```

we have

```
H.χ
```

$\{2\,x + 3\,y,\ 5\,x + 6\,y\}$

```
1
─ χ.H.χ // Expand
2
```

$x^2 + 4\,x\,y + 3\,y^2$

and

```
1
─ χ. (H.χ) + a.χ + b // Expand
2
```

$5 + 7\,x + x^2 - 5\,y + 4\,x\,y + 3\,y^2$

On the other hand, given a quadratic function f, such as

```
f[{x_, y_, z_}] := x² + 2 y² + 3 z² - y z + x y + z - y + 3
```

the Hessian matrix of f can be found as follows:

```
grad[ξ_, χ_] := Table[∂_{χ[[i]]} ξ, {i, Length[χ]}];
hessian[ξ_, χ_] := grad[grad[ξ, χ], χ]

hessian[f[{x, y, z}], {x, y, z}] // MatrixForm
```

$$\begin{pmatrix} 2 & 1 & 0 \\ 1 & 4 & -1 \\ 0 & -1 & 6 \end{pmatrix}$$

The gradient of a quadratic function

$$f(\chi) = \frac{1}{2}\, \chi \cdot H \chi + a \cdot \chi + b$$

is

$$\nabla f(\chi) = H \chi + \mathbf{a}.$$

So it turns out that if if H is invertible, then f has a single critical point $\chi = -H^{-1}\,\mathbf{a}$.

■ Positive Definiteness

A matrix H is said to be **positive definite** if

$$\chi \cdot H \chi > 0 \ \ \text{for all nonzero vectors } \chi.$$

If the matrix H is positive definite, then:

- H is invertible, and

- the function $f(\chi) = \frac{1}{2}\, \chi \cdot H \chi + \mathbf{a} \cdot \chi + \mathbf{b}$ has a single critical point $\chi = H^{-1}\,\mathbf{a}$ at which it attains a global minimum.

The following theorem provides a means for determining whether a matrix is positive definite. Its proof can be found in many linear algebra texts.

⬚ *The matrix H is positive definite if and only if the determinants of all of the* principal minors *of H are positive.*

Any $n \times n$ matrix H has n **principal minors** M_1, M_2, ..., M_n, where M_k is the $k \times k$ matrix obtained by deleting all but the first k rows and columns of H. (Thus $M_n = H$.) For instance if H is the 4×4 matrix given by

```
H = {{1, -1, 0, 0}, {-1, 2, -1, 0}, {0, -1, 3, -1}, {0, 0, -1, 4}};
MatrixForm[H]
```

$$\begin{pmatrix} 1 & -1 & 0 & 0 \\ -1 & 2 & -1 & 0 \\ 0 & -1 & 3 & -1 \\ 0 & 0 & -1 & 4 \end{pmatrix}$$

then the four principal minors of Q are

```
Map[MatrixForm,
  Table[H[[Range[1, k], Range[1, k]]], {k, Length[H]}]]
```

$$\left\{ (1),\ \begin{pmatrix} 1 & -1 \\ -1 & 2 \end{pmatrix},\ \begin{pmatrix} 1 & -1 & 0 \\ -1 & 2 & -1 \\ 0 & -1 & 3 \end{pmatrix},\ \begin{pmatrix} 1 & -1 & 0 & 0 \\ -1 & 2 & -1 & 0 \\ 0 & -1 & 3 & -1 \\ 0 & 0 & -1 & 4 \end{pmatrix} \right\}$$

Our concern here is with the determinants of the principal minors. The following defines a command that will compute a list of the determinants of all the principal minors of a matrix.

```
principalMinorDets[A_ ?MatrixQ] := Map[Det,
  Table[A[[Range[1, k], Range[1, k]]], {k, 1, Length[A]}]]
```

For instance, when we apply this command to the matrix H above, we find

principalMinorDets[H]

$\{1, 1, 2, 7\}$.

which shows that all of the principal minors of H have positive determinants. Therefore we conclude that H is positive definite.

---

1. For the quadratic function

$$f(x, y, z, w) = x^2 + 5y^2 + z^2 + w^2 - yz - 4xy - wz + x - y + 3z - 4w$$

compute the Hessian matrix H and show that H is positive definite. Then find the the minimum value of f.

2. For each of the following functions, compute the Hessian matrix H and determine whether H is positive definite.

    a) $f(x, y) = x^2 + 5y^2 - 7xy + x + 3y + 2$ b) $f(x, y, z) = x^2 + 5y^2 + z^2 - yz - 4xy$

    c) $f(x, y, z) = x^2 + 4y^2 + z^2 - yz - 4xy$ c) $f(w, x, y, z) = w^2 + x^2 + 5y^2 - yz - wz$

3. Let $f(x, y, z) = x^2 + 3y^2 + 5z^2 - 3yz - cxy$. Describe all values of c for which f has a single critical point and a global minimum at $(0, 0, 0)$.

4. A matrix H is said to be **negative definite** if

$$\chi \cdot H\chi < 0 \text{ for all nonzero vectors } \chi.$$

    If the matrix H is negative definite, then:

    - H is invertible, and
    - the function $f(\chi) = \frac{1}{2} \chi \cdot H\chi + a \cdot \chi + b$ has a single critical point $\chi = H^{-1}a$ at which it attains a global *maximum*.

    a) Argue that H is negative definite if and only if $-H$ is positive definite.

    b) Verify that the Hessian matrix of $f(x, y, z) = -x^2 - 2y^2 - 3z^2 - yz + 2xy$ is negative definite. Then find the maximum value of f.

5. If the Hessian matrix H of $f(\chi)$ is invertible and neither positive definite nor negative definite, then f has neither a maximum nor a minimum value at the critical point $\chi = H^{-1}\mathbf{a}$. In this case, the critical point is called a **saddle point**.

    a) Describe all values of c such that $(0, 0)$ is a saddle point of

$$f(x, y) = x^2 + 2y^2 + cxy.$$

    Then plot the graph of $z = f(x, y)$ for one such c.

    b) Describe all values of c such that $(0, 0)$ is a saddle point of $f(x, y) = x^2 - y^2 + cxy$. Then plot the graph of $z = f(x, y)$ for one such c.

c) Check that $(0, 0)$ is a saddle point of $f(x, y) = x^2 - y^2$. Then use `ContourPlot` to plot $f(x, y) = 0$ on the square region $-1 \leq x \leq 1$, $-1 \leq y \leq 1$. Interpret the light and dark regions in the plot. What is the connection with the fact that $(0, 0)$ is a saddle point?

d) Check that $(0, 0, 0)$ is a saddle point of $f(x, y, z) = x^2 + y^2 - z^2$. Then use `Contour-Plot3D` to plot $f(x, y, z) = 0$ on the cube $-1 \leq x \leq 1$, $-1 \leq y \leq 1$, $-1 \leq z \leq 1$. Interpret the regions "inside" and "outside" the quadric surface. What is the connection with the fact that $(0, 0, 0)$ is a saddle point?

■ General Nonlinear Functions

If f is a twice-differentiable function of n variables—i.e., a twice-differentiable function of a vector χ of length n—then the quadratic approximation of f at a point A is given by

$$Q_A(\chi) = f(A) + \nabla f(A) \cdot (\chi - A) + \frac{1}{2}(\chi - A) \cdot \nabla^2 f(A)(\chi - A)$$

where $\nabla^2 f(A)$ is the **Hessian matrix** of f at A, whose ij^{th} entry is

$$\nabla^2 f(A)_{i,j} = \left(\frac{\partial^2 f}{\partial x_i \, \partial x_j} \right)_{\chi = A}.$$

The behavior of f at a critical point A is essentially determined by the behavior of its quadratic approximation at A. More precisely, the following theorem is true:

⬚ Suppose that A is a critical point of f, i.e., $\nabla f(A) = 0$.

• If $\nabla^2 f(A)$ is positive definite, then f has a local minimum at A.

• If $\nabla^2 f(A)$ is negative definite, then f has a local maximum at A.

• If $\nabla^2 f(A)$ is invertible and neither positive nor negative definite, then A is a saddle point.

6. Given the function

$$f(x, y, z) = x \sin x + z^2 \cos(y z) + yz + 2x^2 + xy + y^2 e^{xz}$$

verify that $(0, 0, 0)$ is a critical point. Then show that f has a local minimum there.

7. Consider the function

$$f(x, y, z) = x^4 + y^3 + z^2 + xy + yz + xz.$$

a) Use `NSolve` to locate all of the critical points of f.

b) Compute the Hessian matrix of f at each (real) critical point.

c) At which of critical points is the Hessian matrix of f positive definite?

d) At which of critical points is the Hessian matrix of f negative definite?

e) Are any of the critical points saddle points?

7.9 Least Squares and Curve Fitting

Background material is found in Section 15.7 of Stewart's *Calculus*. (ET:14.7)

A. Suppose that twenty-five fish of a certain species are measured and weighed, resulting in the following list of ordered pairs containing the length and weight measurements (in inches and pounds) for each fish.

data

```
{{9.6, 6.4}, {16.4, 11.7}, {10.4, 6.6}, {11.1, 7.9},
 {11.8, 10.}, {11.7, 8.7}, {18.2, 12.9}, {19.2, 16.2},
 {11.3, 10.3}, {17.7, 15.3}, {11.5, 10.}, {10.3, 8.4},
 {9.5, 8.3}, {8., 7.7}, {5.5, 3.4}, {14.4, 9.5}, {14.9, 12.4},
 {20.9, 17.7}, {20.7, 16.}, {16.5, 14.3}, {5.5, 5.5},
 {19.7, 14.8}, {14.1, 9.5}, {12.1, 8.4}, {16.9, 13.6}}
```

scatterplot = ListPlot[data, PlotRange → {{0, 22}, {0, 18}}]

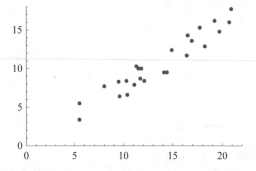

Observing that there seems to be a linear trend in the scatterplot, we would like to determine the equation of the straight line that "*best fits*" the data. So we wish to find values of a and b so that the graph of $m(x) = a x + b$ is as close as possible to our data points. We will do this by minimizing the **least-squares error**, defined here by

$$\mathrm{LSE}(a, b) = \sum_{i=1}^{n} \left(m(x_i) - y_i \right)^2$$

where (x_i, y_i) denotes the ith data point.

---

1. Define the model function

    m[a_, b_][x_] := a x + b

    and then define the least-squares error function

    $$\texttt{LSError[a\_, b\_] = } \sum_{i=1}^{25} \left(\texttt{(m[a, b][data[[i, 1]]] - data[[i, 2]])} \right)^2$$

2. Compute and simplify the partial derivatives of **LSError** with respect to a and b. Then solve for the critical point.

3. Plot the resulting line along with the scatterplot of the data.

4. Predict the weight of such a fish if it were 24 inches long.

B. A biologist is interested in the relationship between air temperature and the presence of *sandgnats* in Savannah, Georgia. Sandgnats are particularly pernicious pests that can practically preclude outdoor activity by humans during certain times of the year. At a certain location prone to sandgnat swarms, the biologist makes temperature measurements and sandgnat counts at sunset on twenty-five different days during the months of April and May. The measurements and a scatterplot are shown below.

data

```
{{80, 141}, {66, 196}, {65, 193}, {67, 203}, {78, 162},
 {78, 155}, {52, 92}, {66, 199}, {58, 165}, {83, 108},
 {49, 44}, {84, 100}, {77, 174}, {56, 131}, {86, 54},
 {70, 188}, {73, 180}, {79, 156}, {77, 183}, {64, 195},
 {76, 193}, {57, 171}, {86, 79}, {67, 225}, {83, 113}}
```

scatterplot = ListPlot[data, PlotRange → {{40, 95}, {0, 230}}]

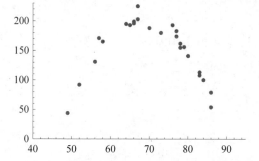

The shape of the scatterplot suggests that there may be a quadratic relationship between temperature and the number of sandgnats. So we wish to find values of a, b and c so that the graph of $m(t) = a\,t^2 + b\,t + c$ is as close as possible to our data points. We will do this by minimizing the **least-squares error**, defined here by

$$\text{LSE}(a, b, c) = \sum_{i=1}^{n}\left(m(t_i) - y_i\right)^2$$

where (t_i, y_i) denotes the ith data point. (Even though the model function is nonlinear, this is still a *linear least-squares problem*, since the gradient of the least-squares error will be a linear function of a, b, and c.)

5. By analogy with problems 1 and 2 above, define the model function and solve for the critical point of the least squares error.

6. Plot the resulting parabola along with the scatterplot of the data.

7. Use the model to estimate the temperature that sandgnats most enjoy.

■ Using `Fit`

Mathematica has a built-in function called `Fit` for solving linear least-squares problems. Find out how `Fit` is used, and then use it to find the model function that best fits the data in:

8. part A above;   9. part B above.

7.10 More Curve Fitting: Nonlinear Least Squares

Suppose that the ordered pairs in the following list represent temperature measurements recorded over a thirty minute period.

    data

    {{0, 50.2}, {1, 50.6}, {2, 51.4}, {3, 52.4}, {4, 53.3}, {5, 53},
     {6, 54.6}, {7, 55.8}, {8, 56.6}, {9, 56.4}, {10, 56.3},
     {11, 56.5}, {12, 55.6}, {13, 56}, {14, 54.6}, {15, 53.9},
     {16, 53.3}, {17, 52.5}, {18, 52.3}, {19, 52.}, {20, 51.3},
     {21, 50.9}, {22, 51.1}, {23, 50.4}, {24, 50.5}, {25, 50.2},
     {26, 50.3}, {27, 50.4}, {28, 50.1}, {29, 50.3}, {30, 50.1}}

Here is a scatterplot of the data:

    ListPlot[data, AxesLabel → {time, temp}]

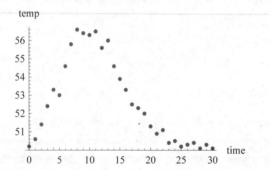

Our goal here is to find a "model function" whose graph closely follows the data points. The appearance of the scatterplot suggests that a function of the form

$$m(t) = a\,t^2\,e^{-b\,t^2} + c$$

might produce reasonable results. Note, for example, the shape of the following graph, where $a = b = c = 1$:

    Plot$\left[t^2\,\text{Exp}\left[-t^2\right] + 1,\ \{t,\ 0,\ 3\},\ \text{Ticks} \to \text{None}\right]$

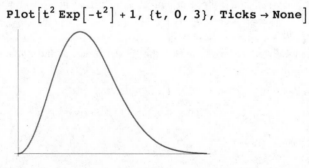

So we wish to find values of a, b, and c so that the graph of $m(t) = a\,t^2\,e^{-b\,t^2} + c$ is as close as possible to our data points. We will do this by minimizing the **least-squares error**, defined here by

$$\text{LSE}(a, b, c) = \sum_{i=1}^{n} \left(m(t_i) - y_i\right)^2$$

where (t_i, y_i) denotes the ith data point. This is a *nonlinear least-squares problem*, since the gradient of the least-squares error will be a nonlinear function of a, b, and c.

1. For the given data and the model function $m(t) = a\,t^2\,e^{-b\,t^2} + c$, find a, b, c that minimize the least squares error. *Suggestion*: First use `FindMinimum` with starting values 0.1, 0.1, and 50 for a, b, and c, respectively.

    After finding a, b, and c, plot the resulting model function. Then `Show` the graph of the model function together with the scatter plot of the data.

A **logistic curve** is the graph of a function of the form

$$f(t) = \frac{a\,b}{a + (b - a)\,e^{-k\,t}}$$

For example,

```
Plot[10 / (1 + 9 Exp[-.1 t]), {t, 0, 80}]
```

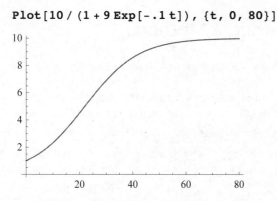

Such functions are sometimes used to model population growth. In this context, a represents the population at time $t = 0$, b represents the limiting population as $t \to \infty$, and k is called the *intrinsic growth rate*.

2. Suppose that the number of deer on a small island is *estimated* annually over a twenty-year period, resulting in the following data.

```
data = {{0, 50}, {1, 62}, {2, 80}, {3, 92}, {4, 120}, {5, 136},
    {6, 165}, {7, 175}, {8, 210}, {9, 230}, {10, 240},
    {11, 260}, {12, 270}, {13, 270}, {14, 275}, {15, 290},
    {16, 290}, {17, 295}, {18, 290}, {19, 290}, {20, 295}};
```

```
ListPlot[data, PlotRange → {0, 300}]
```

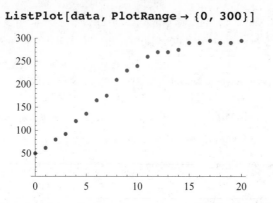

By minimizing the least squares error, find a, b, and k so that the graph of

$$f(t) = \frac{a\,b}{a + (b - a)\,e^{-k\,t}}$$

best fits the data. Use the interpretations of a and b noted above to arrive at initial guesses for those variables. Use an initial guess of 1 for k.

After finding a, b, and k, create a plot that shows the graph of the resulting model function together with the scatterplot of the data.

7.11 The Steepest Descent Method

Background material is found in Sections 15.6 and 15.7 of Stewart's *Calculus*. (ET:14.6,14.7)

The steepest descent method is an iterative method for locating minima of functions of two or more variables. Beginning at an initial guess χ_0, the method updates the current point χ_k with a new point χ_{k+1} that (locally) minimizes the function along the line through the current point in the direction of the negative gradient, thus generating a sequence of points that (we hope) converges to a local "minimizer" of the function.

The method can be stated as follows:

For $k = 0, 1, 2, \ldots$, find the closest point to χ_k on the line $\mathbf{r}(t) = \chi_k - t\,\nabla f(\chi_k)$ at which a local minimum of $\varphi_k(t) = f(\chi_k - t\,\nabla f(\chi_k))$ occurs. Let this point be χ_{k+1}.

Let's illustrate the method with the function

```
f[{x_, y_, z_}] := (x + y - 2)² + (x - z)² + (y - 1)²
```

which has a minimum at $(1, 1, 1)$. We'll need our `grad` command from Chapter 5:

```
grad[ξ_, χ_] := Table[D[ξ, χ[[i]]], {i, Length[χ]}]
```

which we'll use to find the gradient of f:

```
gradf[{x_, y_, z_}] = grad[f[{x, y, z}], {x, y, z}] // Simplify
{2 (-2 + 2 x + y - z), 2 (-3 + x + 2 y), -2 x + 2 z}
```

Each step of the method involves finding t^* to minimize $\varphi_k(t) = f(\chi_k - t\,\nabla f(\chi_k))$ and setting $\chi_{k+1} = \chi_k - t^*\,\nabla f(\chi_k)$. The following function does that job:

```
SDstep[x_] :=
    (x - t gradf[x]) /. FindMinimum[f[x - t gradf[x]], {t, 0}][[2]]
```

Note that by using an initial guess of $t = 0$ in `FindMinimum`, we generally expect (though it's not guaranteed) to find the minimizer closest to the current point. Now, using `Nest-List`, we can easily crank out a list of points in the steepest-descent sequence. Let's use (1.1, 1.2, .9) as a starting point and see what happens.

```
SDseq = NestList[SDstep, {1.1, 1.2, .9}, 5]
```

```
{{1.1, 1.2, 0.9}, {0.944828, 1.04483, 0.962069},
 {0.963926, 1.02095, 0.950133}, {0.965355, 1.01467, 0.964994},
 {0.973298, 1.01682, 0.96514}, {0.975041, 1.00982, 0.973373}}
```

The corresponding function values are

```
Map[f, SDseq]
```

```
{0.17, 0.00241379, 0.000857953, 0.000614379, 0.000447084, 0.000328406}
```

So we see that the method generates a sequence of points—along which f has decreasing values—that seems to be converging (albeit slowly) to the minimizer (1, 1, 1). Let's see what happens if we go further and do 50 steps:

```
SDseq = NestList[SDstep, {1.1, 1.2, .9}, 50]; Last[SDseq]
```

```
{0.999977, 1.00001, 0.999969}
```

The value of f at the last iterate,

```
f[Last[SDseq]]
```

$$3.4643 \times 10^{-10}$$

and the gradient there,

```
gradf[Last[SDseq]]
```

```
{-2.48773 × 10^-6, 0.0000111489, -0.000015283}
```

suggest that the last iterate is close to the exact minimizer.

---

1. The function f defined by

$$f(x, y, z) = (y - x + z)^2 + \left(3 - x - y^2\right)^2 + \left(6 - x - y^2 - z^3\right)^2$$

has a local minimizer near (2, 1, 1.5). Compute twenty steps of the steepest descent method to find an approximation to that minimizer. Compute the values of f and ∇f at the resulting approximation. Compare the result to that given by `FindMinimum`.

2. The function f defined by

$$f(a, b, c) = (a + b \sin c - .92)^2 + (2a + b \sin 2c - 1.16)^2 + (3a + b \sin 3c - .79)^2$$

has a local minimizer near $(a, b, c) = (.3, .7, 1.)$. Compute thirty steps of the steepest descent method to find an approximation to that minimizer. Compute the values of f and ∇f at the resulting approximation. Compare the result to that given by `FindMinimum`.

3. The function f defined by

$$f(x, y) = (x + 3y - 5)^2 + (x - 2y)^2$$

has a global minimizer at $(2, 1)$. Compute twelve steps of the steepest descent method beginning at the point $(0, 1)$. Then create a plot that shows a `ListPlot` of the steepest-descent iterates (with option `PlotJoined→True`) together with a `ContourPlot` of f (with options `Contours→Map[f,SDseq]`, `ContourShading→ False`). What do you observe in the plot?

■ A More Efficient Variation

The steepest descent method as described above wastes a great deal of effort finding highly accurate minimizers along each descent direction, since each of these is only an approximation to the solution we seek. A more efficient variation of the method involves minimizing the quadratic approximation of the restriction of the function to the line of steepest descent at each step. (For *quadratic* functions, this variation in the method changes nothing!) We state this as follows.

For $k = 0, 1, 2, \ldots$

   (i) Let $\varphi_k(t) = f(\chi_k - t\,\nabla f(\chi_k))$.

   (ii) Do *one step* of Newton's Method on the problem of finding a root of $\varphi'_k(t) = 0$ with initial guess $t_0 = 0$, and let t_1 be the resulting value of t.

   (iii) Let $\chi_{k+1} = \chi_k - t_1 \nabla f(\chi_k)$.

If $\varphi_k(t) = f(\chi_k - t\,\nabla f(\chi_k))$, then

$$\varphi'_k(t) = -\nabla f(\chi_k - t\,\nabla f(\chi_k)) \cdot \nabla f(\chi_k),$$

and

$$\varphi''_k(t) = \nabla f(\chi_k - t\,\nabla f(\chi_k)) \cdot H(\chi_k - t\,\nabla f(\chi_k))\,\nabla f(\chi_k).$$

where $H(\chi)$ is the **Hessian matrix** of f at χ, defined for functions of two variables by

$$H(x, y) = \begin{pmatrix} f_{xx}(x, y) & f_{xy}(x, y) \\ f_{yx}(x, y) & f_{yy}(x, y) \end{pmatrix}$$

and for functions of three variables by

$$H(x, y, z) = \begin{pmatrix} f_{xx}(x, y, z) & f_{xy}(x, y, z) & f_{xz}(x, y, z) \\ f_{yx}(x, y, z) & f_{yy}(x, y, z) & f_{yz}(x, y, z) \\ f_{zx}(x, y, z) & f_{zy}(x, y, z) & f_{zz}(x, y, z) \end{pmatrix},$$

and so on. (Please see the remarks about notation at the beginning of Project 7.8.)

Consequently, one step of Newton's Method on the equation $\varphi_k'(t) = 0$ with initial guess $t_0 = 0$ gives

$$t_1 = 0 - \frac{\varphi_k'(0)}{\varphi_k''(0)} = \frac{\nabla f(\chi_k) \cdot \nabla f(\chi_k)}{\nabla f(\chi_k) \cdot H(\chi_k) \nabla f(\chi_k)}.$$

Therefore, the variation on the steepest descent method described above becomes the following simple algorithm:

For $k = 0, 1, 2, \ldots$

Let $\chi_{k+1} = \chi_k - \left(\dfrac{\nabla f(\chi_k) \cdot \nabla f(\chi_k)}{\nabla f(\chi_k) \cdot H(\chi_k) \nabla f(\chi_k)} \right) \nabla f(\chi_k).$

Let's try this out on the function

```
f[{x_, y_, z_}] := e^z (x + y - 2)^2 + (x - z) Sin[x - z] + x^2 (y - 1)^2
```

which has a local minimum value of 0 at (1, 1, 1). First we'll compute the gradient:

```
(gradf[{x_, y_, z_}] =
   grad[f[{x, y, z}], {x, y, z}] // Simplify) // MatrixForm
```

$$\begin{pmatrix} 2\,x\,(-1 + y)^2 + 2\,e^z\,(-2 + x + y) + (x - z)\,\mathrm{Cos}[x - z] + \mathrm{Sin}[x - z] \\ 2\,x^2\,(-1 + y) + 2\,e^z\,(-2 + x + y) \\ e^z\,(-2 + x + y)^2 + (-x + z)\,\mathrm{Cos}[x - z] - \mathrm{Sin}[x - z] \end{pmatrix}$$

and then the Hessian matrix:

```
hessf[{x_, y_, z_}] = grad[gradf[{x, y, z}], {x, y, z}] // Simplify
```

$$\{\{2\,(e^z + (-1 + y)^2) + 2\,\mathrm{Cos}[x - z] + (-x + z)\,\mathrm{Sin}[x - z],$$
$$2\,e^z + 4\,x\,(-1 + y),\ 2\,e^z\,(-2 + x + y) - 2\,\mathrm{Cos}[x - z] + (x - z)\,\mathrm{Sin}[x - z]\},$$
$$\{2\,e^z + 4\,x\,(-1 + y),\ 2\,(e^z + x^2),\ 2\,e^z\,(-2 + x + y)\},$$
$$\{2\,e^z\,(-2 + x + y) - 2\,\mathrm{Cos}[x - z] + (x - z)\,\mathrm{Sin}[x - z],$$
$$2\,e^z\,(-2 + x + y),\ e^z\,(-2 + x + y)^2 + 2\,\mathrm{Cos}[x - z] + (-x + z)\,\mathrm{Sin}[x - z]\}\}$$

One step of the algorithm is encapsulated in the following function:

```
quickSDstep[x_] := With[{gf = gradf[x]}, x - gf.gf/(gf.(hessf[x].gf)) * gf]
```

Before we use this new method, let's see how the original algorithm performs on this problem by computing 100 steps from an initial guess of (1.1, 1.2, .9). For later comparison with the new version of the algorithm, we'll do a `Timing` of the computation.

```
Timing[χ100 = Nest[SDstep, {1.1, 1.2, .9}, 100]]
```

```
{0.227654, {0.999877, 1.00009, 0.999842}}
```

□ (Notice the use of `Nest` here, rather than `NestList`. `Nest` returns only the last iterate.)

So 100 iterates were computed in about 0.23 seconds, and the gradient and function value at the last iterate are

> **gradf[χ100]**
> **f[χ100]**
>
> $\{-0.000116575,\ -8.07949 \times 10^{-6},\ -0.0000695399\}$
>
> 1.23237×10^{-8}

Now let's begin again with an initial guess of (1.1, 1.2, .9) and compute 120 steps of the improved algorithm, again doing a **Timing** of the computation.

> **Timing[χ120 = Nest[quickSDstep, {1.1, 1.2, .9}, 120]]**
>
> $\{0.056027,\ \{0.999996,\ 1.,\ 0.999994\}\}$

So 120 steps were computed in .06 seconds, and the gradient and function value at the last iterate are

> **gradf[χ120]**
> **f[χ120]**
>
> $\{-4.22289 \times 10^{-6},\ -4.28137 \times 10^{-7},\ -2.40255 \times 10^{-6}\}$
>
> 1.50821×10^{-11}

Comparing these numbers with those obtained with the original algorithm, we see that the improved algorithm achieves slightly better accuracy (on *this* problem) in about *one-fourth the time*.

⚠ *Your mileage will vary!*

---

4. The function f defined by

$$f(x, y, z) = (y - x + z)^2 + (3 - x - y^2)^2 + (6 - x - y^2 - z^3)^2$$

   has a local minimizer somewhere near (2, 1, 1.5). Compute 100 steps of the improved steepest-descent algorithm to find an approximation to that minimizer. Compute the values of f and ∇f at the resulting approximation. Compare the result to that in #1 of this project.

For each of the three functions in 5-7, compute the requested number of steps of the original steepest descent algorithm—including a **Timing**—beginning with the specified initial guess. Then, still computing timings, experimentally estimate how many steps of the improved method are needed to achieve the same accuracy. Compare the timings.

5. $f(x, y, z) = \sin\big((x - y + z)^2\big) + \sin^2(x + y - 2) + z^2 x^4;$ $\chi_0 = (.8, 1.3, .2);$ 20   steps.
   ((1, 1, 0) is a local minimizer.)

6. $f(x, y, z) = e^{(x-y)^2} + \sin^2\big(e^y (x^2 + z - 2)\big) + (z^2 - 1)^2;$ $\chi_0 = (.8, 1.3, 1.2);$ 30   steps.
   ((1, 1, 1) is a local minimizer.)

7. $f(x, y, z) = \big(y^2 - z\big)^2 + \big(x^2 - y\big)^2 + \tan^{-1}\big((1.3 - x)^2 + (.8 - z)^2\big);$ $\chi_0 = (1., 1., 1.);$ 30 steps.
   (Gauge the accuracy by the size of the components of the gradient.)

7.12 Quadratic Approximation and Optimization

Background material is found in Sections 15.6 and 15.7 of Stewart's *Calculus*. (ET:14.6-7)

Given a function f of two variables, its quadratic approximation at (a, b) is given by

$$Q_{(a, b)}(x, y) = f(a, b) + f_x(a, b)(x - a) + f_y(a, b)(y - b)$$
$$+ \frac{1}{2}\left(f_{xx}(a, b)(x - a)^2 + 2 f_{xy}(a, b)(x - a)(y - b) + f_{yy}(a, b)(y - b)^2\right)$$

Think of this as the function whose graph is the (elliptic, circular, or hyperbolic) paraboloid that best approximates the surface $z = f(x, y)$ at $(a, b, f(a, b))$. Note that the preceeding formula can be expressed as

$$Q_{(a, b)}(x, y) = f(a, b) + \nabla f(a, b) \cdot \langle x - a, y - b \rangle + \frac{1}{2} \langle x - a, y - b \rangle \left(\nabla^2 f(a, b)\begin{pmatrix} x - a \\ y - b \end{pmatrix}\right)$$

where $\nabla^2 f(a, b)$ is the **Hessian matrix** of f at (a, b), defined by

$$\nabla^2 f(a, b) = \begin{pmatrix} f_{xx}(a, b) & f_{xy}(a, b) \\ f_{yx}(a, b) & f_{yy}(a, b) \end{pmatrix},$$

and the products in the third term are matrix products. More generally, if f is a function of n variables—i.e., a function of a vector χ of length n—then the quadratic approximation of f at the point A is given by

$$Q_A(\chi) = f(A) + \nabla f(A) \cdot (\chi - A) + \frac{1}{2}(\chi - A) \cdot \nabla^2 f(A)(\chi - A).$$

The following is a command for computing a quadratic approximation. Note that it uses the `grad` command from Chapter 4; enter that first. Also note that the composition of `grad` with itself produces the Hessian matrix. (See Exercise 14 after Section 4.4.)

```
quadApprox[ξ_, χ_, a_] := If[Length[χ] == Length[a],
  Module[{subs = Table[χ[[i]] → a[[i]], {i, Length[χ]}]},
   (ξ /. subs) + (Evaluate[grad[ξ, χ]] /. subs).(χ - a) +
    ((χ - a) / 2).(Evaluate[grad[grad[ξ, χ], χ]] /. subs).(χ - a)]]
```

The arguments ξ, χ, and **a** represent the function $f(\chi)$, the vector of variables χ, and the point A, respectively. For example, suppose that

```
f[x_, y_] := Sin[(x - y)²] + Sin[(x + y - 2)]²
```

Then, the quadratic approximation of f at

```
χ := {1, 1}
```

is

```
q[x_, y_] = quadApprox[f[x, y], {x, y}, χ] // Expand
```
$$4 - 4x + 2x^2 - 4y + 2y^2$$

The following plot shows the graph of the function and the quadratic approximation at (1, 1).

```
Show[Plot3D[f[x, y], {x, -1, 2}, {y, 0, 3}],
  Plot3D[q[x, y], {x, 0, 2}, {y, 0, 2}],
  PlotRange → {0, 2}, BoxRatios → Automatic]
```

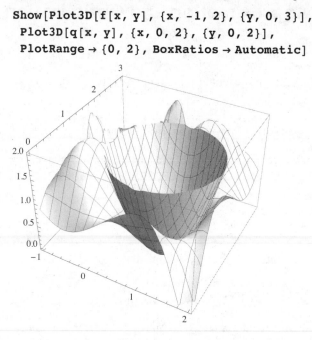

■ Newton's Method for Minimization

Suppose that f is a quadratic function (of any number of variables) that attains a minimum value at a unique point P. In general, the problem of finding this point is simple (though not necessarily easy). This is because the gradient will be a linear function, and so finding the critical point amounts to solving a *linear* system of equations. An iterative method for locating local minimizers of smooth, non-quadratic functions—based on quadratic approximation—takes advantage of this. The method is **Newton's method for minimization** and is described as follows:

> For $k = 0, 1, 2, \ldots$
>
> Find the minimizer of the quadratic approximation of f at χ_k. Let this new point be χ_{k+1}.

This is called Newton's method for minimization because it is equivalent to Newton's method for finding points where the gradient of f is **0**. (See the next project on Newton's method for systems of nonlinear equations.)

The function

$$f[\{x\_, y\_, z\_\}] := \text{Exp}\left[(x - y)^2\right] + \text{Sin}\left[x^2 + z - 2\right]^2 + \left(z^2 - 1\right)^2$$

has a local (and global) minimum at (1, 1, 1). Let's compute a few Newton steps, beginning with an initial guess of

```
χ = {.8, .9, 1.5};
```

The quadratic approximation to f there is

```
quadf = quadApprox[f[{x, y, z}], {x, y, z}, x] // Expand
```

$24.0431 - 8.54551\,x + 3.76691\,x^2 - 0.0040402\,y -$
$2.0605\,x\,y + 1.03025\,y^2 - 32.0671\,z + 3.07538\,x\,z + 12.4611\,z^2$

and its gradient is

```
grad[quadf, {x, y, z}] // Simplify
```

$\{-8.54551 + 7.53382\,x - 2.0605\,y + 3.07538\,z,$
$\quad -0.0040402 - 2.0605\,x + 2.0605\,y, \ -32.0671 + 3.07538\,x + 24.9221\,z\}$

Now we'll use `Solve` to find the point where the gradient of the quadratic approximation is zero.

```
{x, y, z} /. Solve[grad[quadf, {x, y, z}] == 0, {x, y, z}][[1]]
```

$\{0.90158, 0.903541, 1.17544\}$

This is the next iterate. Notice that it is considerably closer to $(1, 1, 1)$ than was the initial guess.

To more efficiently compute several Newton steps, let's encapsulate the process in a function:

```
newtStep3D[x_] :=
  Module[{quadf}, quadf = quadApprox[f[{x, y, z}], {x, y, z}, x];
    {x, y, z} /. Solve[grad[quadf, {x, y, z}] == 0, {x, y, z}][[1]]]
```

Now, the previous calculation can be done by simply entering

```
newtStep3D[{.8, .9, 1.5}]
```

$\{0.90158, 0.903541, 1.17544\}$

With `NestList`, we can compute several iterates quickly and easily:

```
NestList[newtStep3D, {.8, .9, 1.5}, 5]
```

$\{\{0.8, 0.9, 1.5\}, \{0.90158, 0.903541, 1.17544\},$
$\quad \{0.987907, 0.987907, 1.03262\}, \{0.999262, 0.999262, 1.00151\},$
$\quad \{0.999999, 0.999999, 1.\}, \{1., 1., 1.\}\}$

Notice how quickly the iterates converge to the solution.

---

1. The function $f(x, y, z) = (y - x + z)^2 + \left(3 - x - y^2\right)^2 + \left(6 - x - y^2 - z^3\right)^2$ has a local minimizer somewhere near $(2, 1, 1.5)$. Compute four Newton steps to find an approximation to that minimizer. Compute the values of f and ∇f at the resulting approximation. Compare the result to that given by `FindMinimum`.

2. The function $f(x, y, z) = \left(y^2 - z\right)^2 + \left(x^2 - y\right)^2 + \tan^{-1}\!\left((1.3 - x)^2 + (.8 - z)^2\right)$ has a local minimizer somewhere near $(1, 1, 1)$. Compute four Newton steps to find an approximation to that minimizer. Compute the values of f and ∇f at the resulting approximation. Compare the result to that given by `FindMinimum`.

3. The function f defined by $f(x, y) = e^{-\sqrt{x^2+3y^2}} \cos\sqrt{x^2 + y + 2y^2}$ has a local minimizer in the rectangle $1 \le x \le 3, -1 \le y \le 2$. Use a contour plot to obtain a rough estimate of the minimizer. Then use Newton to find the minimizer. (You'll need to modify `newtStep3D` to create an `newtStep2D`.)

7.13 Newton's Method for Nonlinear Equations

Background material is found in Sections 15.6 and 15.7 of Stewart's *Calculus*. (ET:14.6-7)

Recall Newton's Method for finding roots of a function f of one variable:

$$x_{k+1} = x_k - \frac{f(x_k)}{f'(x_k)}, \quad k = 0, 1, 2, \ldots,$$

which was derived by solving the linearized problem at x_k:

$$f(x_k) + f'(x_k)(x - x_k) = 0.$$

The analogous method for a system of equations $\mathbf{F}(\chi) = \mathbf{0}$ where $\mathbf{F}$ is a transformation from $\mathbb{R}^n$ to $\mathbb{R}^n$ is derived in the same way. The linearized problem is

$$\mathbf{F}(\chi_k) + J_\mathbf{F}(\chi_k)(\chi - \chi_k) = 0.$$

where $J_\mathbf{F}(\chi)$ is Jacobian matrix of $\mathbf{F}$ at χ (cf. Section 6.6), and we are thinking of χ, χ_k, and $\mathbf{F}(\chi_k)$ as column vectors. The ij^{th} entry of $J_\mathbf{F}(\chi)$ is the partial derivative of the ith component of $\mathbf{F}$ with respect to the jth variable in χ. The following is a *Mathematica* function that produces the Jacobian matrix:

```
jacobianMat[F_][x_] :=
  Table[D[F[[i]], x[[j]]], {i, Length[F]}, {j, Length[x]}]
```

For example, the Jacobian matrix of the transformation

```
F[{x_, y_, z_}] := {x + 2 y - z, x^2 y + z, x + y z + 2}
```

is

```
jacobianMat[F[{x, y, z}]][{x, y, z}] // MatrixForm
```

$$\begin{pmatrix} 1 & 2 & -1 \\ 2\,x\,y & x^2 & 1 \\ 1 & z & y \end{pmatrix}$$

---

1. Show that Newton's Method for systems of equations takes the form

$$\chi_{k+1} = \chi_k - J_\mathbf{F}(\chi_k)^{-1}\,\mathbf{F}(\chi_k), \quad k = 0, 1, 2, \ldots$$

Because it is more efficient than computing the inverse of the Jacobian matrix at each step, we will restate Newton's method in two steps as:

For $k = 0, 1, 2, \ldots$

     i) Solve $J_{\mathbf{F}}(\chi_k)\,\mathbf{s}_k = -\,\mathbf{F}(\chi_k)$ for $\mathbf{s}_k$.

     ii) Set $\chi_{k+1} = \chi_k + \mathbf{s}_k$.

The following function executes one step of Newton's Method:

```
newtonStep[χk_] :=
 Module[{jF, n = Length[χk], χ = Array[x, Length[χk]]},
  jF = jacobianMat[F[χ]][χ] /. Thread[χ → N[χk]];
  If[Det[jF] == 0,
   χk + Table[.1, {n}], χk + LinearSolve[jF, -F[χk]] ] ]
```

Note that this function checks whether the matrix $J_{\mathbf{F}}(\chi_k)$ has a zero determinant and, if so, uses the step $s_k = (.1, .1, \ldots, .1)$. This avoids errors caused by attempting to solve a linear system with a singular matrix and also will allow an iteration to continue even if we begin at (or later encounter) a point at which $J_{\mathbf{F}}(\chi_k)$ is singular.

Let's use the function $\mathbf{F}$ defined above as an example. Starting with an initial guess of $(0, 0, 0)$, we compute the following sequence:

```
NestList[newtonStep, {0., 0., 0.}, 6]
```

```
{{0., 0., 0.}, {-2., 1., 0.}, {-1.55556, 0.555556, -0.444444},
 {-1.59266, 0.34584, -0.900981}, {-1.66323, 0.349491, -0.964252},
 {-1.66317, 0.348956, -0.965256}, {-1.66317, 0.348956, -0.965257}}
```

As expected, the "value" of $\mathbf{F}$ at the last iterate is essentially $\mathbf{0}$:

```
F[Last[%]]
```

$$\{0., 5.32907 \times 10^{-15}, 1.88738 \times 10^{-14}\}$$

---

2. Use Newton's Method to find a solution of the system
$$x^2 + y^2 + z^2 = 21/64, \quad 8\,x\,z - \sin^2 \pi\, y = 0, \quad 2\,x + 4\,y + 8\,z = 3.$$
Use $(1, 0, 0)$ as the initial guess and compute six iterates.

3. a) Locate a critical point of $f(x, y, z, w) = x + z - x^2\,y - y^2\,z + z\,w + w^2$ by applying Newton's method to its gradient.

 b) This function actually has three (real) critical points. Try to find them all with Newton's method. Compare your results to those given by `NSolve`.

4. The function f defined by
$$f(a, b, c) = (a + b\sin c - .92)^2 + (2\,a + b\sin 2c - 1.16)^2 + (3\,a + b\sin 3c - .79)^2$$

has a local minimizer somewhere near $(a, b, c) = (.3, .7, 1.)$. Use Newton's method to find it. Compute the values of f and $\sqrt{\nabla f \cdot \nabla f}$ at each point in the resulting list of iterates. Are those values strictly decreasing?

5. The function $f(x, y, z) = \left(y^2 - z\right)^2 + \left(x^2 - y\right)^2 + \tan^{-1}\left((1.3 - x)^2 + (.8 - z)^2\right)$ has a local minimizer somewhere near $(1, 1, 1)$. Use Newton's method to find it. Compare the result to that given by `FindMinimum`.

7.14 Constrained Minimization with `FindMinimum`

> Background material is found in Sections 15.7 and 15.8 of Stewart's *Calculus*. (ET:14.7-8)

The `FindMinimum` command is designed for locating local minima and provides a convenient tool for solving unconstrained minimization problems. However, we can use `FindMinimum` to solve certain constrained optimization problems by applying it to an appropriately modified problem.

■ Penalty Functions

For a constrained minimization problem:

$$\text{minimize } f(\chi) \text{ subject to a given constraint on } \chi,$$

the set of all points χ satisfying the given constraint is called the **feasible set**. A **penalty function** is any continuous function $p(\chi)$ that is zero at all points χ in the feasible set and positive elsewhere.

Suppose that we have a minimization problem of the form

$$\text{minimize } f(x, y, z) \text{ subject to } g(x, y, z) = 0.$$

One alternative to the Lagrange multiplers method of Chapter 5 is to replace the constrained problem with a sequence of *unconstrained* problems, each involving an increasingly stringent penalty function. Each penalty function provides a penalty for violation of the original constraint $g(x, y, z) = 0$. One choice of a **penalty function sequence** for this problem is

$$p_n(x, y, z) = n!\, g(x, y, z)^2, \quad n = 1, 2, 3, \ldots$$

The method consists of computing a sequence of points $\chi_n = (x_n, y_n, z_n)$, $n = 1, 2, \ldots$, by minimizing the **modified function**

$$f(x, y, z) + p_n(x, y, z).$$

It can be shown that if this sequence converges, then its limit is a local solution of the original constrained problem. (Can you give an intuitive argument for why this should be true?)

As an example, suppose that we want to find the point closest to the origin on the paraboloid $z = x^2 + y^2 + 3x - y + 1$. The problem then is to

$$\text{minimize } x^2 + y^2 + z^2 \text{ subject to } z - \left(x^2 + y^2 + 3x - y + 1\right) = 0.$$

So first we enter the primary objective function f :

```
f[{x_, y_, z_}] := x^2 + y^2 + z^2
```

and then define a sequence of penalty functions:

```
g[{x_, y_, z_}] = z - (x^2 + y^2 + 3 x - y + 1);
p[n_][{x_, y_, z_}] = n! g[{x, y, z}]^2;
```

Now, with a starting point of

```
x := {0, 0, 1}
```

we compute a new point by minimizing the modified function $f(x, y, z) + p_1(x, y, z)$:

```
FindMinimum[Evaluate[f[{x, y, z}] + p[1][{x, y, z}]],
 {x, x[[1]]}, {y, x[[2]]}, {z, x[[3]]}]
x = {x, y, z} /. %[[2]]
```

```
{0.0968351, {x → -0.25967, y → 0.0865565, z → 0.104678}}
```

```
{-0.25967, 0.0865565, 0.104678}
```

The next new point is found by minimizing $f(x, y, z) + p_2(x, y, z)$:

```
FindMinimum[Evaluate[f[{x, y, z}] + p[2][{x, y, z}]],
 {x, x[[1]]}, {y, x[[2]]}, {z, x[[3]]}]
x = {x, y, z} /. %[[2]]
```

```
{0.1027, {x → -0.27471, y → 0.0915701, z → 0.1121}}
```

```
{-0.27471, 0.0915701, 0.1121}
```

Rather than continue in this "manual" fashion, let's start over after defining

```
PMStep[{j_, {a_, b_, c_}}] :=
 {j + 1, {x, y, z} /. FindMinimum[f[{x, y, z}] + p[j][{x, y, z}],
     {x, a, a + .001}, {y, b, b + .001}, {z, c, c + .001}][[2]]}
```

(The reason for supplying two starting points for each variable is so that `FindMinimum` will not require a symbolic gradient.)

The following computes ten steps of the method, beginning at (1, 1, 1):

```
NestList[PMStep, {1, {1, 1, 1}}, 10]
```

```
{{1, {1, 1, 1}}, {2, {-0.259689, 0.0865329, 0.104628}},
 {3, {-0.27471, 0.09157, 0.1121}},
 {4, {-0.286038, 0.0953462, 0.117812}},
 {5, {-0.290605, 0.0968684, 0.120145}},
 {6, {-0.291855, 0.0972851, 0.120787}},
 {7, {-0.292164, 0.0973897, 0.120794}},
 {8, {-0.292157, 0.0974211, 0.120931}},
 {9, {-0.292165, 0.097421, 0.120931}},
 {10, {-0.292167, 0.097421, 0.120931}},
 {11, {-0.292167, 0.097421, 0.120931}}}
```

This suggests that, to at least four significant figures, a local solution to our constrained minimization problem is

```
soln = Last[%][[2]]
```

```
{-0.292167, 0.097421, 0.120931}
```

Let's check to be sure that the final iterate (approximately) satisfies the constraint:

```
g[soln]
```

-3.32222×10^{-8}

We can also check that the Lagrange multiplier condition is satisfied. This condition says that at the minimizer the gradient of *f* and the gradient of the constraint function must be parallel. This gives the gradient of *f* at the final iterate:

```
grad[f[{x, y, z}], {x, y, z}] /. Thread[{x, y, z} → soln]
```

```
{-0.584333, 0.194842, 0.241862}
```

This is the gradient of *g* at the final iterate:

```
grad[g[{x, y, z}], {x, y, z}] /. Thread[{x, y, z} → soln]
```

```
{-2.41567, 0.805158, 1}
```

This shows that these gradients are (approximately) parallel:

```
%% / %
```

```
{0.241893, 0.241992, 0.241862}
```

■ Inequality Constraints

Suppose that we want to minimize

```
f[{x_, y_, z_}] := -x y^2 z^3 + x^2 y^3 - 3 y z
```

over the unit sphere $x^2 + y^2 + z^2 \le 1$. A sequence of penalty functions for this problem is given by

```
g[{x_, y_, z_}] := 1 - (x^2 + y^2 + z^2);
p[n_][{x_, y_, z_}] := If[x^2 + y^2 + z^2 < 1, 0, 3^n g[{x, y, z}]^2]
```

With a starting point of

```
χ := {0, .5, .5}
```

we compute a new point by minimizing $f(x, y, z) + p_1(x, y, z)$:

```
PMStep[{1, χ}]
```

```
{2, {0.0764968, 0.787721, 0.794346}}
```

This will compute twelve steps of the method, beginning at (.5, .5, .5):

```
NestList[PMStep, {1, {.5, .5, .5}}, 12]
```

```
{{1, {0.5, 0.5, 0.5}}, {2, {0.0764967, 0.78772, 0.794346}},
 {3, {0.0561157, 0.73364, 0.737166}},
 {4, {0.0500085, 0.71473, 0.717523}},
 {5, {0.0480461, 0.708314, 0.710892}},
 {6, {0.0474037, 0.706162, 0.708671}},
 {7, {0.0471886, 0.704791, 0.708581}},
 {8, {0.0471191, 0.704339, 0.708544}},
 {9, {0.0471049, 0.704192, 0.708528}},
 {10, {0.0471239, 0.704148, 0.708515}},
 {11, {0.047174, 0.704136, 0.708505}},
 {12, {0.0472884, 0.704137, 0.708491}},
 {13, {0.0479931, 0.704624, 0.707957}}}
```

It appears as though the solution we've found lies on the boundary of the feasible set (the unit sphere). Let's check:

```
soln = Last[%][[2]]
```

```
{0.0479931, 0.704624, 0.707957}
```

```
√%.%
```

```
1.
```

---

1. For the minimization problem

$$\text{minimize } x^2 + 2\,y^2 + 3\,z^2 \text{ subject to } 1 - 3\,x\,y - 2\,y\,z - x\,z = 0,$$

find a local solution by applying `FindMinimum` to an appropriate sequence of modified functions. Check that the result (approximately) satisfies the constraint and the Lagrange multiplier condition.

2. Given the problem:

$$\text{minimize } x^2 - x\,y\,z \text{ subject } x + 2\,y + 3\,z = 5,$$

find a local solution by applying `FindMinimum` to an appropriate sequence of modified functions. Check that the result (approximately) satisfies the constraint and the Lagrange multiplier condition.

3. Consider the following problem with an inequality constraint:

$$\text{minimize } x^2 - x\,y\,z \text{ subject to } x^2 + 2\,y^2 + 3\,z^2 \le 10.$$

Find a local solution by applying `FindMinimum` to an appropriate sequence of modified functions. Is the solution you found on the boundary of the feasible set or in its interior?

4. Consider this problem involving an inequality constraint:

$$\text{minimize } e^{x^2} + 3\,y + 2\,z^2 + x\,y\,z + 3\,z - 5\,y \text{ subject to } x^2 + y^2 + z^2 \le 2.$$

Find a local solution by applying `FindMinimum` to an appropriate sequence of modified functions. Is the solution you found on the boundary of the feasible set or in its interior?

7.15 3D Coordinate Systems

> Background material is found in Sections 13.7, 15.6, and 16.9 of Stewart's *Calculus*. (ET:*x*–1)

Our purpose here is to survey several different three-dimensional *curvilinear coordinate systems*—including the three with which we are already familiar: rectangular, cylindrical, and spherical coordinates.

General curvilinear coordinates (u, v, w) are the result of applying a differentiable, invertible transformation T given by

$$u = u(x, y, z), \quad v = v(x, y, z), \quad w = w(x, y, z)$$

to the usual (rectangular) Cartesian coordinates (x, y, z). Typically, such coordinate systems are described by means of the inverse transformation T^{-1}, which we'll express as

$$x = x(u, v, w), \quad y = y(u, v, w), \quad z = z(u, v, w).$$

We will refer to any surface on which exactly one of u, v, or w is constant as a **coordinate surface**. Plots of coordinate surfaces often offer a great deal of insight about a given curvilinear coordinate system.

If coordinate surfaces are orthogonal wherever they intersect, then the coordinate system is said to be **orthogonal**. Since coordinate surfaces are just contours (i.e., in this case, *level surfaces*) of functions, we can use the fact that, at any point P,

$$\nabla f(P) \text{ is orthogonal to the contour of } f \text{ that passes through } P,$$

to determine whether a given coordinate system is orthogonal. The result is that $(u(x, y, z), v(x, y, z), w(x, y, z))$ forms an orthogonal coordinate system if

$$\nabla u \cdot \nabla v = \nabla u \cdot \nabla w = \nabla v \cdot \nabla w = 0$$

for all points (x, y, z). Since these gradients are precisely the rows of the transformation's Jacobian matrix $J_T(x, y, z)$, we can say that the coordinate system is orthogonal if its Jacobian matrix $J_T(x, y, z)$ has orthogonal rows. It turns out that $J_T(x, y, z)$ has orthogonal rows if and only if the Jacobian matrix $J_{T^{-1}}(u, v, w)$ of the inverse transformation has orthogonal columns. Consequently, we have the following fact:

☐ *A curvilinear coordinate system* (u, v, w) *is orthogonal if and only if* $J_{T^{-1}}(u, v, w)$ *has orthogonal columns.*

Another useful fact from linear algebra is:

☐ *A matrix M has orthogonal columns if and only if the product* $M^T M$ *is a diagonal matrix.*

(M^T denotes the *transpose* of M, obtained by reflecting the entries of M about its diagonal.)

We also saw in Section 6.6 the *change of variables formula:*

$$\iiint_R f(x, y, z)\, dV = \iiint_S f(x(u, v, w), y(u, v, w), z(u, v, w))\, |\mathcal{J}(u, v, w)|\, du\, dv\, dw,$$

where S is the image of R under the transformation and $\mathcal{J}(u, v, w)$ is the determinant of $J_{T^{-1}}(u, v, w)$. Also recall our *Mathematica* commands for the Jacobian matrix and its determinant:

$$\texttt{jacobianMat3[\Psi\_][\{u\_, v\_, w\_\}]} := \begin{pmatrix} \partial_u \Psi[\![1]\!] & \partial_v \Psi[\![1]\!] & \partial_w \Psi[\![1]\!] \\ \partial_u \Psi[\![2]\!] & \partial_v \Psi[\![2]\!] & \partial_w \Psi[\![2]\!] \\ \partial_u \Psi[\![3]\!] & \partial_v \Psi[\![3]\!] & \partial_w \Psi[\![3]\!] \end{pmatrix};$$

```
jacobianDet3[Ψ_][{u_, v_, w_}] := Det[jacobianMat3[Ψ][{u, v, w}]]
```

■ Rectangular Coordinates

The following is a plot of several coordinate surfaces, which are simply planes that are parallel to the *xy*-, *yz*-, *xz*-planes.

```
Show[
 Map[ContourPlot3D[x == #, {x, -2, 2}, {y, -2, 2}, {z, -1, 1}] &,
  {-1, 0, 1}],
 Map[ContourPlot3D[y == #, {x, -2, 2}, {y, -2, 2}, {z, -1, 1}] &,
  {-1, 0, 1}],
 ContourPlot3D[z == 0, {x, -2, 2}, {y, -2, 2}, {z, -1, 1}],
 BoxRatios → Automatic]
```

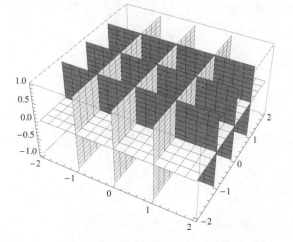

■ (Circular) Cylindrical Coordinates

The (inverse) cylindrical-coordinate transformation is given by

$$x = r \cos\theta, \quad y = r \sin\theta, \quad z = z,$$

where $r \geq 0$, $0 \leq \theta \leq 2\pi$, and $-\infty \leq z \leq \infty$. The following provides a *Mathematica* definition of this transformation.

```
cylToRect[{r_, θ_, z_}] := {r Cos[θ], r Sin[θ], z}
```

A plot of a few coordinate surfaces is constructed as follows.

```
Show[Map[ParametricPlot3D[cylToRect[{#, θ, z}],
    {θ, 0, 2 π}, {z, -1, 2}] &, {.75, 1.5}],
  Map[ParametricPlot3D[cylToRect[{r, #, z}],
    {r, 0, 2}, {z, -1, 2}] &, π / 4 Range[0, 7] ],
  Map[ParametricPlot3D[cylToRect[{r, θ, #}], {r, 0, 2},
    {θ, 0, 2 π}] &, {0, 1}], PlotRange → All]
```

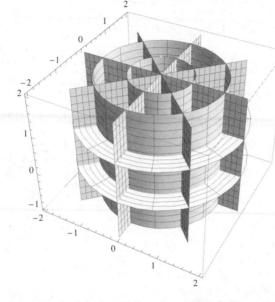

Let's verify the (obvious?) fact that this coordinate system is orthogonal. First we compute the Jacobian matrix $J_{T^{-1}}(r, \theta, z)$:

```
jacobianMat3[cylToRect[{r, θ, z}]][{r, θ, z}] // MatrixForm
```

$$\begin{pmatrix} \text{Cos}[\theta] & -r\,\text{Sin}[\theta] & 0 \\ \text{Sin}[\theta] & r\,\text{Cos}[\theta] & 0 \\ 0 & 0 & 1 \end{pmatrix}$$

We can verify the orthogonality of the coordinate system by observing that

```
Transpose[%].% // Simplify // MatrixForm
```

$$\begin{pmatrix} 1 & 0 & 0 \\ 0 & r^2 & 0 \\ 0 & 0 & 1 \end{pmatrix}$$

is a diagonal matrix. Furthermore, the familiar "extra factor" in cylindrical coordinate integrals is

```
jacobianDet3[cylToRect[{r, θ, z}]][{r, θ, z}] // Simplify
```

r

■ Parabolic Cylindrical Coordinates

Parabolic cylindrical coordinates are described by

$$x = \left(u^2 - v^2\right)/2, \quad y = u\,v, \quad z = z,$$

where $-\infty \le u \le \infty$, $v \ge 0$, and $-\infty \le z \le \infty$. A *Mathematica* definition of this transformation is

```
parCylToRect[{u_, v_, z_}] := {(u² - v²) / 2, u v, z}
```

1. Create a plot showing parabolic cylindrical coordinate surfaces corresponding to:

   i) $u = 1$, $-2 \le v \le 2$, $-1 \le z \le 2$

   ii) $-2 \le u \le 2$; $v = .5$, 1.2; $-1 \le z \le 2$

   iii) $-2 \le u \le 2$; $-2 \le v \le 2$; $z = 0$, 1

2. a) Compute $J_{T^{-1}}(u, v, z)$ and check that its columns are orthogonal.

   b) Compute the Jacobian determinan.

   c) Create a plot of the solid described by $0 \le u \le 1$, $0 \le v \le 1$, $0 \le z \le 1$, and compute its volume.

■ Spherical Coordinates

Recall the defining relations for spherical coordinates:

$$x = \rho \sin \phi \cos \theta, \quad y = \rho \sin \phi \sin \theta, \quad \text{and} \quad z = \rho \cos \phi$$

where $\rho \ge 0$, $0 \le \phi \le \pi$, and $0 \le \theta \le 2\pi$. A *Mathematica* definition of this transformation is

```
sphrToRect[{ρ_, θ_, φ_}] := {ρ Sin[φ] Cos[θ], ρ Sin[φ] Sin[θ], ρ Cos[φ]}
```

3. Create a plot showing spherical coordinate surfaces corresponding to:

   i) $\rho = 1.1$;   ii) $0 \le \rho \le 2$, $\theta = \pi/3$, $-\pi/6$;   iii) $0 \le \rho \le 2$, $\phi = \pi/6$, $\pi/4$, $3\pi/4$

4. a) Compute $J_{T^{-1}}(\rho, \theta, \phi)$ and its determinant, and check that its columns are orthogonal.

   b) Create a plot of the solid described by $0 \le \rho \le 1$, $0 \le \theta \le 1$, $0 \le \phi \le 1$, and compute its volume.

■ Paraboloidal Coordinates

Paraboloidal coordinates are described by

$$x = u\,v \cos \theta, \quad y = u\,v \sin \theta, \quad \text{and} \quad z = \left(u^2 - v^2\right)/2$$

where u, $v \ge 0$ and $0 \le \phi \le 2\pi$. A *Mathematica* definition of this transformation is:

```
parabToRect[{u_, v_, θ_}] := {u v Cos[θ], u v Sin[θ], (u² - v²) / 2}
```

5. Create a plot showing paraboloidal coordinate surfaces corresponding to:

   i) $u = 1$, $0 \le v \le 2$ ii) $v = 1$, $0 \le u \le 2$ iii) $-2 \le u \le 2$, $-2 \le v \le 2$, $\phi = 0, \pi/2$

6. a) Compute $J_{T^{-1}}(u, v, \theta)$ and its determinant, and check that its columns are orthogonal.

   b) Create a plot of the solid described by $0 \le u \le 1$, $0 \le v \le 1$, $0 \le \theta \le 1$, and compute its volume.

■ Prolate Spheroidal Coordinates

Prolate spheroidal coordinates are described by

$$x = \sinh u \sin \phi \cos \theta, \ \ y = \sinh u \sin \phi \sin \theta, \ \text{and} \ \ z = \cosh u \cos \phi$$

where $u \ge 0, 0 \le \theta \le 2\pi$, and $0 \le \phi \le \pi$. A *Mathematica* definition of this transformation is:

```
prSphrToRect[{u_, θ_, φ_}] :=
     {Sinh[u] Sin[φ] Cos[θ], Sinh[u] Sin[φ] Sin[θ], Cosh[u] Cos[φ]}
```

7. Create a plot showing prolate spheroidal coordinate surfaces corresponding to:

   i) $u = 1/2$ ii) $0 \le u \le 1$, $\theta = -\pi/12, \pi/6$ iii) $\phi = \pi/4, \pi/3, 3\pi/4$

8. a) Compute $J_{T^{-1}}(u, \theta, \phi)$ and its determinant, and check that its columns are orthogonal.

   b) Create a plot of the solid described by $0 \le u \le 1$, $0 \le \theta \le 1$, $0 \le \phi \le 1$, and compute its volume.

■ Oblate Spheroidal Coordinates

Oblate spheroidal coordinates are described by

$$x = \cosh u \sin \phi \cos \theta, \ \ y = \cosh u \sin \phi \sin \theta, \ \text{and} \ \ z = \sinh u \cos \phi$$

where $u \ge 0$, $0 \le \theta \le 2\pi$, and $-\pi/2 \le \phi \le \pi/2$. A *Mathematica* definition of this transformation is

```
obSphrToRect[{u_, θ_, φ_}] :=
     {Cosh[u] Cos[φ] Cos[θ], Cosh[u] Cos[φ] Sin[θ], Sinh[u] Sin[φ]}
```

9. Create a plot showing prolate spheroidal coordinate surfaces corresponding to:

   i) $u = 1/2$ ii) $0 \le u \le 1$, $\theta = -\pi/12, \pi/6$ iii) $\phi = \pi/4, \pi/3, 3\pi/4$

10. a) Compute $J_{T^{-1}}(u, \theta, \phi)$ and its determinant, and check that its columns are orthogonal.

    b) Create a plot of the solid described by $0 \le u \le 1$, $0 \le \theta \le 1$, $0 \le \phi \le 1$, and compute its volume.

7.16 Rolling Marbles on Surfaces

> Background and related material are found in Sections 14.4 and 15.6 of Stewart's *Calculus*.

Let's assume for simplicity a gravitational constant of 1, so that the downward force of gravity on the marble (i.e., its *weight*) is numerically equal to its mass m. Also, we'll assume that the marble will remain on the surface at all times and that no forces other than gravity act on it—in particular, there is no friction (for now).

At any point $(x, y, f(x, y))$, the components of the gravitational force vector $\langle 0, 0, -m \rangle$ in the directions of the vectors $\langle 1, 0, f_x(x, y) \rangle$ and $\langle 0, 1, f_y(x, y) \rangle$ are

$$\frac{-m\, f_x(x, y)}{\sqrt{1 + f_x(x, y)^2}} \quad \text{and} \quad \frac{-m\, f_y(x, y)}{\sqrt{1 + f_y(x, y)^2}} \,.$$

Consequently, by Newton's Second Law, the first two components of the particle's acceleration vector will be

$$x''(t) = \frac{-f_x(x, y)}{\sqrt{1 + f_x(x, y)^2}} \quad \text{and} \quad y''(t) = \frac{-f_y(x, y)}{\sqrt{1 + f_y(x, y)^2}}.$$

These differential equations, together with $z(t) = f(x(t), y(t))$, will govern the motion of the marble. *Mathematica*'s `NDSolve` command will compute numerical approximations to solutions of these equations.

Let's start with a simple example, using the quadratic function

```
f[x_, y_] := (x^2 + y^2) / 2
```

We'll first make a semi-transparent surface:

```
surf = ParametricPlot3D[
    {r Cos[θ], r Sin[θ], f[r Cos[θ], r Sin[θ]]}, {r, 0, 3},
    {θ, 0, 2 π}, PlotStyle → Opacity[.5], Mesh → None]
```

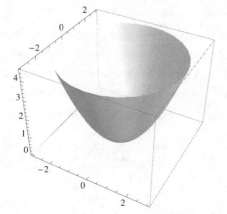

Next we'll first compute partial derivatives

```
fx[x_, y_] = ∂x f[x, y]; fy[x_, y_] = ∂y f[x, y];
```

and then create a list containing the appropriate equations. Notice the (first two) coordinates x_0, y_0 of the initial point and u_0, v_0 of the initial velocity.

```
diffeqs[{x0_, y0_}, {u0_, v0_}] :=

  {x"[t] == -fx[x[t], y[t]] / √(1 + fx[x[t], y[t]]²) ,

   y"[t] == -fy[x[t], y[t]] / √(1 + fy[x[t], y[t]]²) ,

   x[0] == x0, x'[0] == u0, y[0] == y0, y'[0] == v0}
```

The following commands compute the solution and produce a parametric plot of the marble's path.

```
endTime = 17; initPt = {0, 2.5}; initVel = {1, 0};
soln = Flatten[
  NDSolve[diffeqs[initPt, initVel], {x, y}, {t, 0, endTime}]];
r[t_] = {x[t], y[t], f[x[t], y[t]]} /. soln;
Show[ParametricPlot3D[r[t] + {0, 0, .05}, {t, 0, endTime},
  PlotStyle → AbsoluteThickness[1]], surf, PlotRange → All]
```

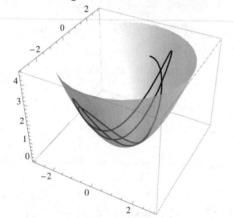

1. Reproduce the example plot given above. Then do the following.

    a) Leave $x_0 = 0$, $u_0 = 1$, and $v_0 = 0$ fixed and vary y_0 to experimentally approximate the value that will cause the marble's path to be a nearly horizontal, simple closed path.

    b) Leave $x_0 = 0$, $y_0 = 2.5$, and $v_0 = 0$ fixed and vary u_0 to experimentally approximate the value that will cause the marble's path to be a nearly horizontal, simple closed path.

2. If we assume a friction force that is proportional to the velocity of the marble, then the acceleration equations become

$$x''(t) = \frac{-f_x(x, y)}{\sqrt{1 + f_x(x, y)^2}} - \delta x'(t) \text{ and } y''(t) = \frac{-f_y(x, y)}{\sqrt{1 + f_y(x, y)^2}} - \delta y'(t)$$

where δ is a positive constant.

a) Modify the procedure described above to include these friction terms. Then, using $\delta = 0.1$, examine the effect of friction with $x_0 = 0$, $y_0 = 2.5$, $u_0 = 1$, and $v_0 = 0$.

b) Using $x_0 = 0$, $y_0 = 2.5$, $v_0 = 0$, and the value of u_0 that you found in #1b, examine the effect of friction on the path of the marble. Use $\delta = .02$, 0.1, and 0.5.

3. Now it's time to play *hole-in-one!* Your "putting green" is the surface $z = f(x, y)$, over the square region $-5 \le x \le 5$, $-5 \le y \le 5$, where

$$f[x\_, y\_] := 1 - x/73 + y/91 - e^{-30\,(x^2+y^2)}$$

Use the following to create the wireframe surface. (Note that the surface has a slight but significant slope.

```
surf = Show[Plot3D[f[x, y], {x, -5, 5}, {y, -5, 5},
    PlotPoints → 40, PlotStyle → Darker[Green], Mesh → None,
    BoxRatios → Automatic, Axes → False, Boxed → False]]
```

a) Using a friction coefficient of $\delta = 0.05$, putt from the point $(4, 3)$ and experimentally determine an initial velocity vector $\langle u_0, v_0 \rangle$ that will put your ball in the "hole." Your ball must not only go in the hole, it must *stay* in the hole, so continue the simulation of the path long enough to be sure it does. Try this "shot" for starters:

```
endTime := 30; initPt = {4, 3}; initVel = {-.4, -.2};
```

b) Putt from each of the points $(-4, 1)$ and $(-4, -2)$. Then try the uphill putt from $(3, -4)$.

7.17 Balancing a Region

Background material is found in Section 15.6 and 16.7 of Stewart's *Calculus*. (ET:14.6,15.7)

Suppose that a three-dimensional solid occupies a bounded region R in three-dimensional space. Suppose also that this solid can be moved about in space without changing its shape. For example, if R is defined as

$$R = \left\{ (x, y, z) \mid x^2 + y^2 + z^2 \le 1 \right\},$$

then we can think of a ball of radius 1 occupying R and imagine rolling or throwing the ball. If we place this spherical ball on a horizontal surface, it will balance on the point of contact, which can be any point on the surface of the ball. This will not be true for other choices of R. For example, consider the *ellipsoid*

$$R = \left\{ (x, y, z) \;\middle|\; x^2 + \frac{y^2}{4} + \frac{z^2}{3} \le 1 \right\}.$$

If the corresponding ellipsoidal ball that occupies R is placed on a horizontal surface, then it will not balance on the point of contact unless that point corresponds to one of the special *balancing points* on the boundary of R.

1. Describe all the balancing points of the ellipsoid R described above.

In general, such a problem can be solved in two steps: First find the center of mass of R. Then find all points on the boundary of R at which the line normal to the boundary passes through the center of mass.

2. Consider the region

$$R = \left\{(x, y, z) \mid x^2 + y^2 - 1 \le z \le 8 - 2x^2 - 2y^2\right\}.$$

Find the set of all balancing points of R.

3. Repeat #2 for

$$R = \left\{(x, y, z) \mid x^2 + y^2 - 1 \le z \le 8 - 2x^2 - y^2\right\}.$$

A balancing point P is *stable* if when the solid is placed on a horizontal surface with point of contact close to P, the solid might wobble or "rock" some before coming to rest, but its point of contact with the surface will stay close to P.

4. Consider the balancing points in each of problems 1-3. In each case, which of the balancing points will be stable? Give reasons for your answers.

7.18 Reflecting Points

Some related material is found in Section 15.6 of Stewart's *Calculus*. (ET:14.6)

Suppose that C is a plane curve and that P is a given point on C. We say that P is a reflecting point if there is another point Q on C so that the normal line to C at Q passes through P. If you imagine that the curve is a mirror, then a source of light at P, shining toward Q, would bounce directly back to P.

1. Find all the reflecting points on each of the following curves.

   a) $y = x^2$ b) $y = x^3$ c) $y = x^3 - x$

Similarly, if S is a surface in three-space, then a point P on S is a reflecting point if there is another point Q on S such that the line normal to the surface at Q passes through P.

2. a) Give an example of a surface in three-space that has no reflecting points.

   b) Give an example of a surface in three-space for which *every* point is a reflecting point.

3. Let $f(x, y) = x^2 + y^2$. Describe the set of points (a, b) in the xy-plane for which $P = (a, b, f(a, b))$ is a reflecting point on the surface $z = f(x, y)$.

4. Repeat #3 with $f(x, y) = x^2 y$.

5. a) Repeat #3 with $f(x, y) = \left(\varepsilon + x^2 + y^2\right)^{-1}$ for ε equal to 1, 1/2, and 1/4.

   b) For what (positive) values of ε does the surface $z = \left(\varepsilon + x^2 + y^2\right)^{-1}$ have a nonempty set of reflecting points?

7.19 2D Space Exploration

Background and related material is found in Sections 14.4 and 17.1 of *Calculus*. (ET:13.4,16.1)

Suppose that a particle moves in space under the influence of a conservative force field $\mathbf{F} = -\nabla f$. At time $t = 0$ the particle is located at a point χ_0 and has a velocity vector $\mathbf{V}_0$. What is the subsequent path of the particle?

Let's assume for simplicity that the mass of the particle is 1 unit. By Newton's Second Law, the particle's acceleration vector will be

$$\chi''(t) = -\nabla f(\chi(t)).$$

This differential equation will govern the motion of the particle.

Let's start with a simple two-dimensional example, using the quadratic function

```
f[x_, y_] := (2 x^2 + y^2) / 3
```

A contour plot is a plot of *equipotential curves*:

```
equipots = ContourPlot[f[x, y],
    {x, -2, 2}, {y, -2, 2}, ContourShading → False]
```

Now let's compute partials:

```
fx[x_, y_] = ∂x f[x, y]; fy[x_, y_] = ∂y f[x, y];
```

and plot the force field:

```
<< VectorFieldPlots`;
field = VectorFieldPlot[-{fx[x, y], fy[x, y]},
    {x, -2, 2}, {y, -2, 2}, PlotPoints → 9]
```

Assuming that the particle has mass 1, the differential equations and initial conditions are set up as follows.

```
diffeqs[{x0_, y0_}, {u0_, v0_}] :=
  {x"[t] == -fx[x[t], y[t]], y"[t] == -fy[x[t], y[t]],
   x[0] == x0, x'[0] == u0, y[0] == y0, y'[0] == v0}
```

Now we're ready to plot the path. We'll superimpose the path, the force field, and the equipotential curves. Using an initial point $(1, 1)$ and initial velocity vector $\langle -1, -.5 \rangle$, we see the following picture.

```
endTime = 20; initPt = {1, 1}; initVel = {-1, -.5};
soln =
  First[NDSolve[diffeqs[initPt, initVel], {x, y}, {t, 0, endTime}]];
Show[ParametricPlot[Evaluate[{x[t], y[t]} /. soln],
  {t, 0, endTime}, PlotStyle → Thickness[.007]], field, equipots,
  AspectRatio → Automatic, Axes → False, Frame → False, PlotRange → All]
```

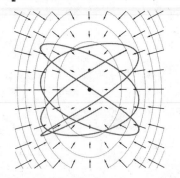

▽ Suppose that you are the captain of a spacecraft that is approaching a small planet in two-dimensional space. The planet has a radius of 0.5 units and has a gravitational force field with the potential function

```
φ[x_, y_] := -1 / √(x² + y²)
```

$$\phi[x\_, y\_] := -1 \Big/ \sqrt{x^2 + y^2}$$

where $(0, 0)$ is the location of the planet's center. Your mission is to enter an orbit around the planet that keeps your spacecraft within 5 units of the planet's center. You must approach the planet by shutting off your thrusters as you pass through the point $(2, 2)$ (at time $t = 0$) with a velocity vector $\langle u_0, 0 \rangle$. Your approach to the planet will thus be determined by u_0.

1. Set up the appropriate differential equations and initial conditions for your spacecraft's flight toward the planet from the point $(2, 2)$. Then with an initial velocity vector $\langle -1, 0 \rangle$, create a plot with options AspectRatio→Automatic and PlotRange→ {{-5,5},{-5,5}} that shows the trajectory for $0 \le t \le 10$ and an image of the planet (Graphics[Disk[{0,0},.5]]). Did you achieve orbit around the planet? (Don't bother plotting the contours or the vector field.)

2. Experimentally determine the range of values of u_0 that result in a successful mission. *Crashing into the planet is not a successful mission!*

Suppose that a planet with radius 0.5 units is located at the point $(2, 0)$ and a moon with radius 0.1 units is located at the point $(-2, 0)$. The combined gravitational attraction of both the planet and the moon produces this potential function:

```
φ[x_, y_] := - 1 / √( (x - 2)² + y² ) - .1 / √( (x + 2)² + y² )
```

Your spacecraft is in a counterclockwise orbit around the planet that carries you through the point $(3, 0)$. Your mission is to briefly apply thrusters as you pass through $(3, 0)$ in order to enter a periodic "figure-8" orbit about both the planet and its moon. Your velocity vector at $(3, 0)$ (at time $t = 0$) will be $\langle 0, v_0 \rangle$. Your subsequent flight will thus be determined by v_0.

3. Set up the appropriate differential equations and initial conditions for your spacecraft's flight from the point $(3, 0)$. First try an initial velocity vector $\langle 0, 1 \rangle$. Create a plot with options `AspectRatio→Automatic` and `PlotRange→{{-4,4},{-3,3}}` that shows the trajectory for $0 \le t \le 40$ along with images of the planet and its moon.

4. Experiment with values of v_0 between 1.2 and 1.3, plotting trajectories for $0 \le t \le 40$, and find the value of v_0 that achieves your mission.

7.20 Streamlines I: Velocity Fields and Steady Flow

Background material is found in Section 17.1 of Stewart's *Calculus*. (ET:16.1)

A continuous vector field determines a family of **flow lines**. These are smooth curves that are everywhere tangent to the vector field.

In the context of fluid flow, the vector field of interest is the *velocity field* of the fluid at each point in its domain. When the velocity field is independent of time, the flow is said to be a *steady flow,* and flow lines are usually called **streamlines**. So imagine a steady flow of fluid particles moving in two dimensions in such a way that velocity (independently of time) is a function of location (x, y) given by

$$V(x, y) = \langle u(x, y), v(x, y) \rangle.$$

For example, suppose that

```
V[x_, y_] := {x - 2 y, x + y}
```

We can plot the velocity field as follows:

```
<< VectorFieldPlots`
vecfield = VectorFieldPlot[V[x, y],
  {x, -3, 3}, {y, -2, 2}, ScaleFactor → .5]
```

How do we plot the resulting streamlines? First we'll define the appropriate list of differential equations and initial conditions:

```
diffeqs[{x0_, y0_}] = {x'[t] == V[x[t], y[t]][[1]],
    y'[t] == V[x[t], y[t]][[2]], x[0] == x0, y[0] == y0}
```

$$\{x'[t] == x[t] - 2\,y[t],\ y'[t] == x[t] + y[t],\ x[0] == x0,\ y[0] == y0\}$$

Then, using NDSolve, we can plot a single streamline along with the vector field as follows.

```
endTime = 4; initPt = {-.07, 0}; soln =
Flatten[NDSolve[diffeqs[initPt], {x, y}, {t, 0, endTime}]];

Show[ParametricPlot[Evaluate[{x[t], y[t]} /. soln], {t, 0, endTime},
    PlotStyle → Thickness[.007]], vecfield, PlotRange → All]
```

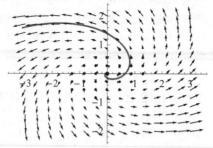

We will want to plot several streamlines at once, each one beginning at a different starting point. So we'll define a list of starting points such as

```
startingpoints := {{0, -.05}, {-.05, 0}, {0, .05}, {.05, 0}}
```

and in order that we can use different final times for different streamlines, we'll also define a list of corresponding final times. In this particular situation, we might start by using the same final time for each streamline by defining

```
stopTimes := {4, 4, 4, 4, 4, 4}
```

and then, after seeing the plot, adjust each number to our liking. With all this in place, the following creates the desired plot:

```
Show[vecfield,
  Table[soln = Flatten[NDSolve[diffeqs[startingpoints[[i]]],
      {x, y}, {t, 0, stopTimes[[i]]}]];
    ParametricPlot[Evaluate[{x[t], y[t]} /. soln],
      {t, 0, stopTimes[[i]]}, PlotStyle → Thickness[.007],
      PlotRange → All], {i, Length[startingpoints]}]]
```

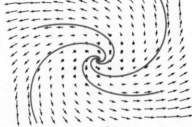

1. a) Plot the vector field

$$V[x\_,\ y\_] := \left\{ \frac{y}{x^2 + y^2},\ \frac{x}{x^2 + y^2} \right\}$$

on the square region described by $-2 \le x \le 2$, $-2 \le y \le 2$. Use `PlotPoints→10`.

b) Plot streamlines with initial points at $(-2, 1.9)$ and $(2, -1.9)$, for $0 \le t \le 8$ in each case, Showing the result superimposed over the vector field. (You may want to plot the streamlines in color and/or with increased thickness.) What kind of curves do the streamlines appear to be? Prove your conjecture.

2. a) Plot the vector field

$$V[x\_,\ y\_] := \left\{ 2 + \frac{x}{x^2 + y^2},\ \frac{y}{x^2 + y^2} \right\}$$

on the square described by $-2 \le x \le 2$, $-2 \le y \le 2$. Use `PlotPoints→10`.

b) Plot streamlines with initial points and final times given as follows, Showing all the streamlines superimposed over the vector field.

`startingpoints := {{-2, -1}, {-2, -.5}, {-2, -.1},`
`   {-2, .1}, {-2, .5}, {-2, 1}, {.01, -.1}, {.01, .1}}`

`stopTimes := {2, 2, 2, 2, 2, 2, 1, 1}`

c) Adjust the entries in the `stopTimes` list so that all of the streamlines reach the edge of the plot.

7.21 Streamlines II: Incompressible Potential Flow

Background material is found in Sections 17.1 and 17.5 of Stewart's *Calculus*. (ET:16.1,16.5)

Given a steady two-dimensional fluid flow with velocity field $V(x, y) = \langle u(x, y), v(x, y) \rangle$, its **rotation** is defined as $\omega = \text{rot}\ V = v_x - u_y$.

For a steady three-dimensional fluid flow with velocity field V, the **vorticity** of the flow is defined to be the curl of the velocity

$$\omega = \nabla \times V.$$

In either case, if $\omega = 0$ (or $\omega = 0$) at all points within the flow, then the flow is said to be **irrotational**.

To imagine an irrotational flow, imagine one in which a *very* short straw will move without rotating; i.e., without changing its angle of orientation with respect to any coordinate axis. (For example, if the straw is parallel to the y-axis at time $t = 0$, then it will remain parallel to the y-axis for all subsequent times. In reality though, no straw is short enough for this to happen!)

It turns out that irrotationality is a necessary and sufficient condition for a flow to be a **potential flow**, that is, a flow whose velocity field V is the gradient of some potential function $\phi(\chi)$.

1. Check that the 2D flow whose velocity field is

$$V(x, y) = \langle 2\,x\,y - 2\,y^2, \ x^2 - 4\,x\,y \rangle$$

   is irrotational and therefore a potential flow. Find the potential function ϕ. Plot the flow field on the square $-2 \le x \le 2, -2 \le y \le 2$.

2. Check that the 3D flow whose velocity field is

$$V(x, y) = \langle 3\,x^2\,y + 3\,z^2, \ x^3 - z, \ -y + 6\,x\,z \rangle$$

   is irrotational and therefore a potential flow. Find the potential function ϕ. Plot the flow field on the cube $-2 \le x \le 2, -2 \le y \le 2, -2 \le z \le 2..$

The flow of an **incompressible** fluid with velocity field V must satisfy the *continuity equation*

$$\mathrm{div}\ V = 0.$$

(The flow of an incompressible fluid is said to be *divergence-free*. Water is an example of an essentially incompressible fluid. Air is an example of a fluid that is compressible.)

Thus, a steady **incompressible** *and* **irrotational flow** is one whose velocity field V satisfies:

$$\mathrm{div}\ V = 0 \text{ and rot } V = 0 \ \text{ for a 2D flow;}$$

$$\mathrm{div}\ V = 0 \text{ and curl } V = \mathbf{0} \ \text{ for a 3D flow.}$$

Irrotationality implies that V has a potential function ϕ. This, together with the continuity equation, implies that ϕ satisfies

$$\nabla \cdot V = \nabla \cdot \nabla \phi = 0;$$

that is, *the velocity potential ϕ satisfies Laplace's equation*:

$$\Delta \phi = 0.$$

Moreover, the gradient of any function ϕ that satisfies Laplace's equation satisfies the conditions to be the velocity field of an incompressible, irrotational fluid flow.

---

3. Show that

$$\phi[\mathbf{x\_, y\_}] := 2\,\mathbf{x} + \frac{1}{2}\,\mathbf{Log}\left[\mathbf{x}^2 + \mathbf{y}^2\right]$$

   satisfies Laplace's equation for all $(x, y) \ne (0, 0)$. Show that the associated velocity field is that of #2 in Project 7.20. Therefore, the streamlines plotted there are those of an incompressible potential flow. (This potential function models a uniform left-to-right flow superimposed over a "source" located at the origin. Imagine a river flowing over a spring.)

4. a) Show that the vector field

$$V[\mathbf{x\_, y\_}] := \left\{ 1 - \frac{\mathbf{x}^2 - \mathbf{y}^2}{\left(\mathbf{x}^2 + \mathbf{y}^2\right)^2}, \ -\frac{2\,\mathbf{x}\,\mathbf{y}}{\left(\mathbf{x}^2 + \mathbf{y}^2\right)^2} \right\}$$

   satisfies div $V = 0$ and rot $V = 0$ for $(x, y) \ne (0, 0)$ and is therefore the velocity field of an incompressible potential flow in any region that does not contain $(0, 0)$.

b) Find a potential function and verify that it satisfies Laplace's equation.

c) Plot the vector field

$$W[x\_, y\_] := If[x^2 + y^2 > 1, V[x, y], \{0, 0\}]$$

on the square region $-2 \le x \le 2$, $-2 \le y \le 2$. (Note that for $x^2 + y^2 > 1$ this is the same vector field in part (a).)

d) Plot streamlines of the velocity field $V[x,y]$ corresponding to:

```
startpoints :=
  {{-2, .1}, {-2, -.1}, {-2, .5}, {-2, -.5}, {-2, -1}, {-2, 1}}

stopTimes := {4, 4, 4, 4, 4, 4}
```

Then display the streamlines and Graphics[Disk[{0,0},1] together with the vector field. Explain—in terms of the movement of fluid particles—what this plot might represent.

e) Adjust the entries in the stopTimes list so that all of the streamlines just reach the edge of the plot.

Appendix
Mathematica Basics

This chapter is an introduction to *Mathematica*. We briefly describe many of the most important and basic elements of *Mathematica* and discuss a few of the more common technical issues related to using *Mathematica*. Since our primary goal is to use *Mathematica* to help us understand calculus, you should not initially spend a great amount of time pouring over the details of this chapter, except as directed by your professor. Simply familiarize yourself with what's here, and refer back to it later as needed.

A.1 Getting Started

Any new user of *Mathematica* must understand several basic facts concerning the user interface, syntax, and the various types of objects that one encounters in using *Mathematica*. This section is a cursory look at some of these fundamentals.

■ The *Mathematica* "Front End"

When you start up *Mathematica*, the first thing you see is a window displaying the contents of a "**notebook**." This window is displayed by *Mathematica*'s **front end**. The front end is the interface between you and the *Mathematica* **kernel**, which does the computations. The following is a typical (simple) notebook in a front end window.

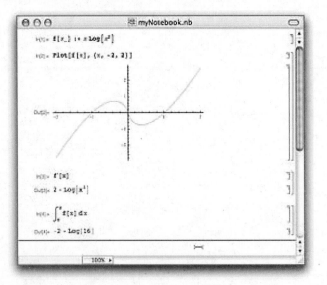

A *Mathematica* notebook is composed of **cells**. On the right side of the window you see **cell brackets**. Each cell in the notebook shown above is either an *input cell*, an *output cell*, or a *graphics cell*. There are several other kinds of cells. Some of these are *text*, *title*, and *section*.

Also notice the horizontal line near the bottom of the window. This indicates the insertion point for the next new cell. To enter a command into a notebook, simply begin typing. The default cell type is *input*. When you're done typing, just press **shift-return** (on a Macintosh, you can also use the "**enter**" key.) To evaluate an existing input cell, simply click anywhere inside the cell (or on the cell bracket) and press **shift-return** (or **enter**).

To create a cell *between* two existing cells, move your cursor over one of the cells toward the other until the "I-beam" cursor becomes horizontal. Then click, and a horizontal line will appear, indicating the desired insertion point. To delete a cell, click on its bracket and then choose Clear from the Edit menu or simply press the DEL key

Palettes. You can enter mathematical expressions so that they appear essentially the same as you would write them on paper or see them in your textbook. For example, to define the

function $f(x) = \sqrt{x^2 + 1}$, we could use the "Input Form"

```
f[x_] := Sqrt[x^2 + 1]
```

or we could use "Standard Form":

```
f[x_] := √(x² + 1)
```

There is a vast set of keystroke combinations for typing such expressions. However, at first, you will probably want to take advantage of one or more or the standard *palettes* that are available. The image to the right shows the BasicInput palette. Clicking on one of the palette's buttons places the corresponding character/expression at the current input location. This particular palette probably appears by default when you start *Mathematica*, but if not, you can access it or any of the other palettes through your File menu as indicated below.

■ The Documentation Center

Mathematica's Documentation Center may be accessed by selecting Documentation Center from the Help menu. Among the wealth of information available through the Documentation Center are descriptions of all of *Mathematica*'s built-in functions, including examples of their use and links to related tutorials. You can also enter commands from within the Documentation Center. (Whatever you change there will not be saved.)

The Documentation Center provides a great deal of tutorial material. If you're a beginner, we suggest that you peruse the tutorials found by entering each of the following in the Documentation Center search field.

> tutorial/UsingTheMathematicaSystemOverview
>
> tutorial/InputAndOutputInNotebooksOverview
>
> tutorial/BuildingUpCalculationsOverview
>
> tutorial/EnteringTwoDimensionalExpressionsOverview
>
> tutorial/GraphicsAndSoundOverview

(Hopefully future updates to the Documentation Center will provide a more convenient way to access these and other tutorials.)

■ Basic Calculations

Mathematica allows multiplication to be indicated in three ways. Expressions separated by a space are multiplied. (Note that the system automatically replaces the space with ×.)

217 × 5713

1 239 721

An asterisk between expressions indicates multiplication.

321 * 5.479

1758.76

When there is no ambiguity, juxtaposed expressions (without spaces) are understood and multiplied.

2x

2 x

More than one command can be given in one input cell. A single input cell may consist of two or more lines. A new line within the current cells is obtained by pressing "return." Commands on the same line within a cell must be separated by semicolons. *The output of any command that ends with a semicolon is not displayed.*

a = 17 / 13 + 211 / 93;
b = 23 a; c = (a + b) / 51

$\dfrac{34\,592}{20\,553}$

The percent sign % refers to the last output (not necessarily the preceding cell).

3 / 17 + 1 / 5

$\dfrac{32}{85}$

%²

$\dfrac{1024}{7225}$

If the last command on any line is not followed by a semicolon, its result *is* displayed. This effect is very handy for showing intermediate steps in a calculation. The following computes $\frac{25!}{3! \times 22!} (.1)^3 (.9)^{22}$ (a *binomial probability*).

25 !
% / (3 ! × 22 !)
% * .1³ .9²²

15 511 210 043 330 985 984 000 000

2300

0.226497

⬚ You should avoid use of the percent sign as much as possible—especially in separate cells. It is *far* better to give names to results and to use those names in subsequent calculations.

■ Parentheses, Brackets, and Braces

The syntax of *Mathematica* is absolutely strict and consistent (and quite simple once you get used to it). For that reason, there are some differences between *Mathematica*'s syntax and the often inconsistent and sometimes ambiguous mathematical notation that we're all used to. For example:

Parentheses are used *only* for grouping expressions.

> **x (x + 2)^2**
>
> x (2 + x)^2

Brackets are used *only* to enclose the argument(s) of a function.

> **Cos[π / 3]**
>
> $\frac{1}{2}$

Braces are used *only* to enclosed the elements of a *list* (which might represent a set, an ordered pair, or even a matrix).

> **{1, 2, 3, 4}**
>
> {1, 2, 3, 4}

Consequently, *Mathematica* does *not* understand what you intend by entering any of these expressions, for example:

> **[x + y (1 - y)]^2**
> Syntax::tsntxi : "[x + y (1 − y)]" is incomplete; more input is needed.
> Syntax::sntxi : Incomplete expression; more input is needed.
>
> **(1, 2)**
> Syntax::sntxf : "(" cannot be followed by "1, 2)".
> Syntax::tsntxi : "1, 2" is incomplete; more input is needed.
>
> **Sin (π)**
>
> π Sin

In these first two instances, we were lucky to get an error message. But in the last one, *Mathematica* simply multiplied the expressions π and Sin—with no complaint at all!

■ Symbolic vs. Numerical Computation

Computations are typically done symbolically (and therefore *exactly*), unless we request otherwise.

> **123 $/ \sqrt{768}$**
>
> $\frac{41 \sqrt{3}}{16}$

One way to obtain a numerical result is to use the numerical evaluation function, N.

$$N\left[123 / \sqrt{768}\,\right]$$

4.43838

We also get a numerical result if any of the numbers in the expression are made numerical by use of a decimal point.

$$123. / \sqrt{768}$$

4.43838

Unless we *cause* a numerical result, *Mathematica* typically returns an *exact* form, which in many cases is identical to the expression entered.

$$\textbf{Cos}\left[\frac{\pi}{12}\right]$$

$$\frac{1+\sqrt{3}}{2\sqrt{2}}$$

$$\textbf{Log[2]}$$

Log[2]

■ Names and Capitalization; Basic Functions

All built-in *Mathematica* objects—functions, constants, options, etc.—have full names that begin with a capital letter (or in the case of certain "global" parameters, a dollar sign followed by a capital letter).

$$\textbf{Sin}[\pi / 3]$$

$$\frac{\sqrt{3}}{2}$$

$$\textbf{PrimeQ}[22\,801\,763\,489]$$

True

$$\textbf{Solve}\left[x^2 + x - 12 == 0,\ x\right]$$

$$\{\{x \rightarrow -4\},\ \{x \rightarrow 3\}\}$$

These full names are used internally by *Mathematica*, even when it is far more natural for us to use a symbolic form such as $x + 7$. FullForm lets us see the internal representation of an expression.

$$\textbf{FullForm}[x + 7]$$

Plus[7, x]

$$\textbf{FullForm}\left[x == x^2\right]$$

Equal[x, Power[x, 2]]

When the name of a built-in *Mathematica* object is made of two or more words, all of the component words are capitalized. Some typical *Mathematica*-style names are `FindRoot`, `PlotRange`, `AspectRatio`, `NestList`, etc. In almost all cases the component words are spelled out in full.

All of the familiar "elementary functions" are built-in. In some cases—if you remember to capitalize the first letter and to use brackets instead of braces—you would guess correctly how to use one of those functions. For example,

> `Sin[π / 12]`
>
> $$\frac{-1+\sqrt{3}}{2\sqrt{2}}$$

There are a few things in this regard that should be pointed out. First, the inverse trigonometric functions use the "arc-function" convention:

> `ArcTan[1]`
>
> $$\frac{\pi}{4}$$
>
> `ArcCos[1 / 2]`
>
> $$\frac{\pi}{3}$$

Also, the natural logarithm is `Log`, not `Ln`.

> `Log[E]`
>
> 1

■ Algebraic Manipulation

Mathematica is an example of a type of software system that is often called a *computer algebra system*. In addition to numerical computations, a computer algebra system also does *symbolic computation* including the manipulation of algebraic expressions. *Mathematica* has a number of functions for this purpose. Among these are `Expand`, `Factor`, `Together`, and `Apart`.

> `Expand[ (x + 5)³ (2 x - 1)² ]`
>
> $125 - 425\,x + 215\,x^2 + 241\,x^3 + 56\,x^4 + 4\,x^5$
>
> `Factor[ x³ + 2 x² - 5 x - 6 ]`
>
> $(-2 + x)\,(1 + x)\,(3 + x)$
>
> `Together[ x + 2/(x² + 1) ]`
>
> $$\frac{2 + x + x^3}{1 + x^2}$$
>
> `Apart[ x/(x² + 3 x + 2) ]`
>
> $$-\frac{1}{1+x} + \frac{2}{2+x}$$

Notice that *Mathematica* does not automatically simplify algebraic expressions:

x (3 - x) - 5 x² + (x - 1) (2 x + 3)

(3 - x) x - 5 x² + (-1 + x) (3 + 2 x)

Simplify can be used for this purpose.

Simplify[x (3 - x) - 5 x² + (x - 1) (2 x + 3)]

-3 + 4 x - 4 x²

■ Plotting Graphs: An Introduction to Options

Mathematica is extremely good at creating graphics to help us analyze problems. We will be primarily interested in graphing functions of one variable. This is done with Plot.

The function $f(x) = \sin(\pi x (3 - x))$ is graphed on the interval $0 \le x \le 3$ as follows.

Plot[Sin[π x (3 - x)], {x, 0, 3}]

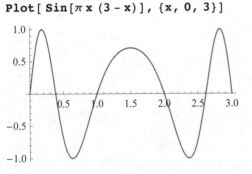

Notice that two *arguments* are provided to Plot. The first is our function in the form of an *expression*, and the second is a *list* with three members, specifying (i) the name of the variable, (ii-iii) the left and right endpoints of the interval.

There are numerous ways that we could have affected the appearance of the plot by specifying **options**. Among the options for Plot are PlotRange, Ticks, AxesLabel, AspectRatio, and PlotStyle.

The following creates a plot with labelled axes with no tick marks. Note that the arrow character is typed as [ESC]->[ESC]. (Actually, just -> will do.)

Plot[Sin[π x (3 - x)], {x, 0, 3},
Ticks → {{1, 2, 3}, {-1, 1}}, AxesLabel → {x, y}]

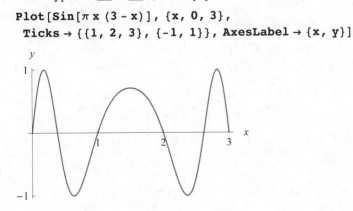

Notice that the following plot chops off high and low parts of the curve.

$$\texttt{Plot}\left[\frac{\texttt{Sin}\left[2\,\texttt{x}^2\right]}{\texttt{x}^2+1},\ \{\texttt{x},\ 0,\ 10\}\right]$$

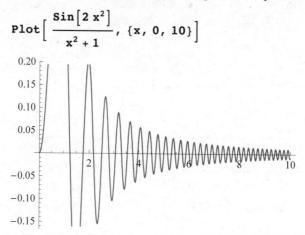

This can be cured with the `PlotRange` option.

$$\texttt{Plot}\left[\frac{\texttt{Sin}\left[2\,\texttt{x}^2\right]}{\texttt{x}^2+1},\ \{\texttt{x},\ 0,\ 10\},\ \texttt{PlotRange}\rightarrow\texttt{All}\right]$$

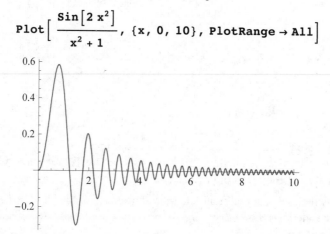

Without our specifying `AspectRatio → Automatic`, the following semicircle would be stretched vertically.

$$\texttt{Plot}\left[\sqrt{1-\texttt{x}^2}\,,\ \{\texttt{x},\ -1,\ 1\},\ \texttt{AspectRatio}\rightarrow\texttt{Automatic}\right]$$

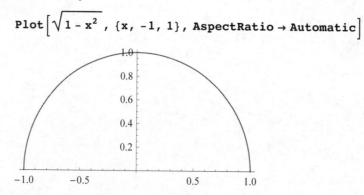

To plot more than one function at once, we give `Plot` a *list* of functions.

$$\text{Plot}\big[\{x\,e^{-x/2}\,\text{Cos}\,[\pi\,x]\,,\ x\,e^{-x/2}\,\text{Sin}\,[\pi\,x]\}\,,\ \{x,\ 0,\ 12\}\big]$$

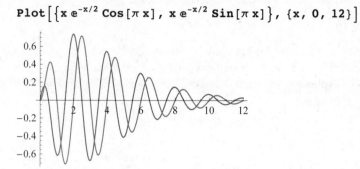

When plotting multiple functions, it is often desirable to plot them with different styles. The `PlotStyle` option lets us do that.

$$\text{Plot}\big[\{\sqrt[3]{x}\,,\ x,\ x^3\}\,,\ \{x,\ 0,\ 2\}\,,\ \text{PlotRange} \to \{0,\ 1.6\}\,,$$
$$\text{PlotStyle} \to \{\{\text{Dashing}\,[\{.02\}]\}\,,\ \{\text{Red}\}\,,\ \{\text{Thickness}\,[.007]\}\}\big]$$

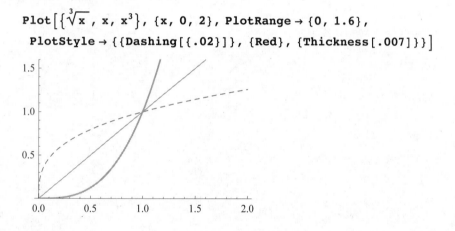

■ Variables

As we mentioned earlier, all built-in *Mathematica* objects begin with an upper-case letter. For that reason, it is usually a good idea to use variable names that begin with a lower-case letter. In this manual we will loosely follow that convention. It is also good practice to give meaningful names to variables and never to make assignments to single letter variables.

⧠ Assignments to variables are remembered by *Mathematica* (for the duration of one "kernel session") until the variable is "cleared." *This is probably the single most important thing to remember when you run into difficulties using Mathematica.* (We will have more to say on this in Section A.8.)

```
piSixths = N[π / 6]
```

0.523599

```
Clear[piSixths]; piSixths
```

0.523599

```
piSixths
```

Assignments in Mathematica are made in two ways: (i) with a single equal sign, or (ii) with a colon followed by an equal sign. For simple assignments such as

radius := $\sqrt{10.}\,/\,\pi$

it makes little difference which method is used. In this particular case, the consequence of using := is that `radius` has not yet been computed. (Also notice that no output is produced when := is used.) The evaluation has been delayed until we cause it to be done—for example, by entering

radius

1.78412

A better example to illustrate delayed evaluation with := is as follows. If we assign a plot to a variable with :=, then no plot is created. The variable is assigned *the command itself*, not the result.

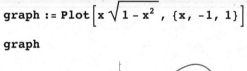

graph

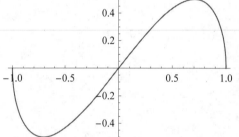

If we assign a plot to a variable with =, then the plot is created and the variable is assigned the resulting *graphics object*.

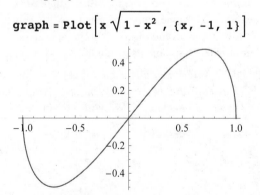

We will address these issues again in Section A.2. (See also Exercise 9 in this section.)

■ Tips and Shortcuts

We end this quick tour of *Mathematica* with a few tips and shortcuts with respect to typing expressions.

■ Entering Exponents, Radicals, and Fractions

To enter an exponent or superscript, press CTRL-[6]. (Note that this is analogous to the SHIFT-[6] caret (^), which is used for exponentiation in InputForm.) To leave the resulting exponent "box," press [→] or CTRL-SPACE.

To enter a subscript, press CTRL-[-]. (This is analogous to the SHIFT-[-] underscore character (_), which is used to create subscripts in the T$_E$X typesetting language.) To leave the resulting subscript "box," press [→] or CTRL-SPACE.

To enter an expression involving a square root, press CTRL-[2]. To leave the resulting square root box, press [→] or CTRL-SPACE.

To enter a fraction, press CTRL-[/]. To move from the numerator box to the denominator box, press TAB To exit the fraction, press [→] or CTRL-SPACE.

■ Greek Letters and Other Special Characters

Many special characters and symbols can be typed easily by pressing the ESC key before and after typing some easily remembered standard character(s). For instance, to type the Greek letter α (alpha), just type ESC a ESC or ESC alpha ESC. Other Greek letters can be typed similarly.

Other shortcuts for common special characters include:

ESC ee ESC produces the constant e, the base of the natural logarithm.

ESC ii ESC produces the imaginary number $i = \sqrt{-1}$.

ESC int ESC produces an integral sign $\int$.

ESC pd ESC produces a derivative operator ∂.

ESC inf ESC produces an infinity symbol ∞ .

Of course, you may prefer to use the buttons on the BasicInput palette.

We should also point out that the familiar numbers denoted by e, i, and π can be entered as such or as E, I, and Pi, respectively.

```
{e == E, i == I, π == Pi}
```

```
{True, True, True}
```

Be careful to notice, however, that an ordinary e is not the same as e, and an ordinary i is not the same as i.

◆ Exercises

1. Compute both an exact and a numerical value for each of the following numbers.

    a) $\left(23^3 - 3\,(117 - 48)^2\right)\Big/\sqrt{7^5 - 5^7}$ b) $\cos\frac{319\,\pi}{12}$ c) $\frac{83!}{111!}$ d) $\ln 2981$

2. Use `Simplify` on each of the following.

    a) $\ln\!\left(2e^5\right)$ b) $1 + \cos 2x$ c) $\dfrac{x + (x\,(x - 1))^3 - 4}{x^2 + x - 6}$

3. Factor each of these polynomials:

    a) $6x^3 + 47x^2 + 71x - 70$ b) $12x^6 - 56x^5 + 100x^4 - 80x^3 + 20x^2 + 8x - 4$

4. Plot the function $f(x) = \dfrac{\sin(x^3)}{x^3}$ on the interval $0 \le x \le 1$ with:

    a) no options   b) `PlotRange→All`

    c) `PlotRange→All, AspectRatio→Automatic`

5. Create a plot containing the graphs of $y = x^2$ and $y = x^5$ over $0 \le x \le 2$ with:

    a) no options   b) `PlotRange→All`   c) `PlotRange→{0,2}`

    d) `PlotStyle→{Red,Blue}`   e) `PlotStyle→{{Red,Thick},{Blue,Thick}}`

6. Look up each of the following functions in the Documentation Center, and then plot them on the indicated interval.

    a) `Floor`, $0 \le x \le 10$ b) `PrimePi`, $0 \le x \le 100$

7. Use the Documentation Center to learn what `RandomReal[]` and `RandomInteger[]` do. Then enter and the following and explain the output:

```
GraphicsRow[{Plot[RandomReal[] x, {x, 0, 1}],
    Plot[RandomInteger[] x, {x, 0, 1}]}]
```

8. `RandomReal` and `RandomInteger` provide a good illustrations of the difference between using `=` and using `:=` in an assignment. Enter each of the following several times. Describe and explain the difference in the results.

```
r = RandomInteger[100]; Table[r, {10}]
```

    and

```
r := RandomInteger[100]; Table[r, {10}]
```

A.2 Functions

■ Defining Functions

■ **Blank**

When defining a function, it is essential to follow each argument by a `Blank` (or "underscore"). Also, recall that the arguments of a function are enclosed by brackets. For example, we would define the function $f(x) = x^3 - 2$ by entering

> `f[x_] := x³ - 2 x`

Then we can evaluate the function at any number

> `f[3]`

> 21

or plot its graph:

> `Plot[f[x], {x, -2, 2}]`

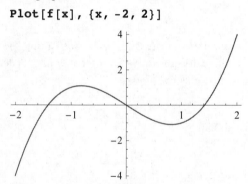

⚠ Note that the `Blank` appears next to x *only on the left side of the expression.* Also, when defining a function with more than one argument, a `Blank` must follow each one. For instance:

> `g[x_, y_, z_] := x + y + z`
> `g[1, 2, 3]`
> 6

■ **Set (=) versus SetDelayed (:=)**

Definitions of functions—and assignments of expressions to variables in general—can be made using either "equal" or "colon-equal." The difference between these two ways is described by the full *Mathematica* names of the = and := symbols, which are `Set` and `SetDelayed`.

When an assignment is made using `Set`, any calculations that are indicated on the right side are done as the assignment is entered. When an assignment is made using `SetDelayed`, any calculations that are indicated on the right side are delayed until the defined expression is used.

In many cases, such as in the definitions of f and g above, it makes no difference which is used. To see a simple example that indicates the importance of using Set rather than SetDelayed, let's suppose we want to define $f(x)$ to be the derivative of $(x+1)\cos x$. If we enter

    f[x_] := ∂ₓ((x + 1) Cos[x])

notice what happens when we try to evaluate $f(2)$:

    f[2]
    General::ivar : 2 is not a valid variable. ≫
    ∂₂ (3 Cos[2])

However, if we enter

    f[x_] = ∂ₓ((x + 1) Cos[x])
    Cos[x] - (1 + x) Sin[x]

then f works the way we want it to:

    f[2]
    Cos[2] - 3 Sin[2]

When is it important to use SetDelayed rather than Set? Here's a simple example: Suppose that we want to plot the graph of $\sin kx$ for several values of k and that we decide to define a function as follows to create each of the desired plots.

    plotSin[k_] := Plot[Sin[k x], {x, 0, 2 π}]

For instance, with $k = 2$, we get the following graph:

    plotSin[2]

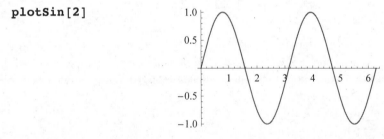

But what would have happened if we used = rather than := in the definition of plotSin? *Mathematica* would attempted to plot the graph of $\sin kx$ immediately when the definition is entered. This doesn't work because k has no numerical value, and an empty plot results.

    plotSin[k_] = Plot[Sin[k x], {x, 0, 2 π}, PlotRange → All]

The empty plot that was produced and assigned to `plotSin[k_]` will now be the "value" of `plotsin[k]` for any k; for instance:

plotSin[2]

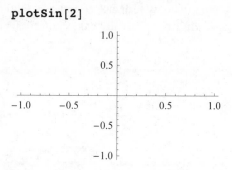

■ Applying Functions with @ and //

Suppose that we define a simple function such as

f[x_] := x (x - 1)

Naturally, we could evaluate this function at, say, $x = 3$, by entering

f[3]

6

There are two other ways to do the same thing. We can evaluate *f at* 3 by entering

f@3

-1

On the other hand, we can apply *f* to 3 like this:

3 // f

6

We will use this *postfix* method of function application frequently, often for the purpose of applying either `Simplify` to some expression or `N` to some numerical calculation. For example, we will use the following style when doing a symbolic calculation:

2 x + x (5 x + 1) // Simplify

x (3 + 5 x)

When doing an exact numerical calculation, we will commonly use a style that is similar but displays the exact value followed by the numerical value:

$$\sqrt{585} \, \big/ \, 33$$

% // N

$$\frac{\sqrt{65}}{11}$$

0.732933

■ Piecewise-defined Functions

Mathematica has two primary logical functions that we can use to enter definitions of piecewise- defined functions. These are If and Which. If usually works best for functions with two pieces, such as

$$f(x) = \begin{cases} x, & \text{if } x \le 1; \\ x-2, & \text{if } x > 1. \end{cases}$$

This function can be entered and plotted as follows.

```
f[x_] := If[x ≤ 1, x, x - 2]
Plot[f[x], {x, -1, 3}]
```

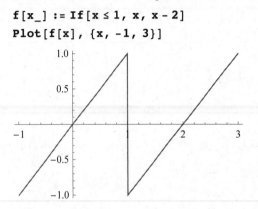

More complicated functions are better handled with Which. For example, the function

$$f(x) = \begin{cases} 1, & \text{if } x \le -1 \\ -x, & \text{if } -1 < x \le 1 \\ -1, & \text{if } x > 1 \end{cases}$$

can be entered and plotted as follows.

```
f[x_] := Which[x ≤ -1, 1, -1 < x ≤ 1, -x, x > 1, -1]

Plot[f[x], {x, -3, 3}]
```

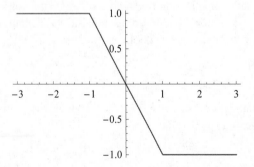

■ Piecewise

Piecewise was new in *Mathematica* 5.1 and is similar to Which. While Which must be supplied with an alternating sequence of conditions and values, *i.e.*,

$$\text{Which}[\text{cond}_1, \text{val}_1, \text{cond}_2, \text{val}_2, \dots],$$

`Piecewise` takes a single list containing value-condition *pairs*, as in

$$\text{Piecewise}[\,\{\{val_1, cond_1\}, \{val_2, cond_2\}, \ldots\}\,]$$

For example, the function

$$f(x) = \begin{cases} 1, & \text{if } x \le -1 \\ -x, & \text{if } -1 < x \le 1 \\ -1, & \text{if } x > 1 \end{cases}$$

can be input as follows. Notice the automatic display of the output in "left-bracket form."

```
f[x_] = Piecewise[{{1, x ≤ -1}, {-x, -1 < x ≤ 1}, {-1, 1 < x}}]
```

$$\begin{cases} 1 & x \le -1 \\ -x & -1 < x \le 1 \\ -1 & 1 < x \end{cases}$$

It is also possible to enter such a definition in the same left-bracket form as follows: Press
ESC pw ESC followed by CTRL RET. Then you'll have a template like this:

$$\begin{cases} \square & \square \\ \square & \square \end{cases}$$

Pressing CTRL RET again will add another row.

▪ Abs, Floor, and Mod

There are a handful of built-in piecewise-defined functions, including `Abs`, `Floor`, and
`Mod`, whose graphs are shown below.

```
GraphicsRow[
  {Plot[Abs[x], {x, -1, 1}], Plot[Floor[x], {x, -2, 3}],
  Plot[Mod[x, 1], {x, 0, 3}]}]
```

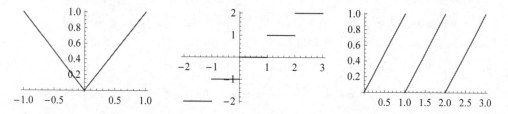

The following functions are built with `Abs`.

```
GraphicsRow[{Plot[Abs[Cos[x] - .5], {x, 0, 4 π}],
  Plot[Abs[Abs[x] - 1], {x, -3, 3}]}]
```

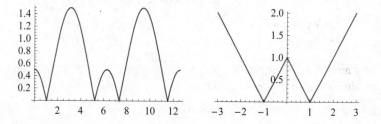

Floor helped create each of these *step functions*:

$$\texttt{GraphicsRow}\big[\{\texttt{Plot}\big[(-1)^{\texttt{Floor[x]}},\ \{x,\ -2,\ 3\}\big],$$
$$\texttt{Plot}[\texttt{Sin}[\pi\,\texttt{Floor}[6\,x]\ /\ 6],\ \{x,\ 0,\ 2\}]\}\big]$$

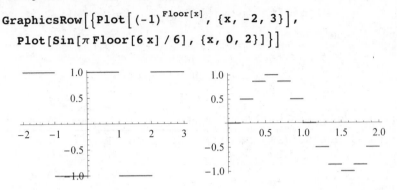

A *periodic function* based on a piece of any graph can be constructed with Mod.

$$\texttt{GraphicsRow}[\{\texttt{Plot}[\texttt{Sin}[\pi\,\texttt{Mod}[x,\ 1]\ /\ 2],\ \{x,\ 0,\ 3\}],$$
$$\texttt{Plot}[\texttt{Abs}[\texttt{Mod}[x,\ 2]\ -\ 1],\ \{x,\ -4,\ 4\}]\}]$$

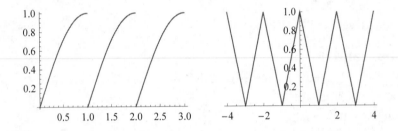

◆ Exercises

1. Enter definitions for each of $f(x) = x\,(x-2)^2$, $g(x) = x - 3$, and $h(x) = x^2 - 1$ and then compute and simplify each of the following:

   a) $f(g(h(x)))$ b) $h(g(f(x)))$ c) $f(h(g(x)))$

2. Enter a definition for $f(x) = \left(1 + x^2\right)^{-1}$. Then create a plot (over $-3 \le x \le 3$) showing the graphs of:

   a) $f(x)$, $f(x-1)$, and $f(x+1)$ b) $f(x)$, $f(2x)$, and $f(8x)$

3. Using If, define and plot each of a) $f(x) = \begin{cases} -x, & \text{if } x < 0 \\ x^2 - 1, & \text{if } x \ge 0 \end{cases}$ b) $f(x) = \begin{cases} 1, & \text{if } x \le \pi/2 \\ \sin x, & \text{if } x > \pi/2 \end{cases}$

4. Using Which, define and plot each of

   a) $f(x) = \begin{cases} 1, & \text{if } x < 0 \\ 1 - x^2, & \text{if } 0 \le x \le 2 \\ -3, & \text{if } x > 2 \end{cases}$ b) $f(x) = \begin{cases} x + 1, & \text{if } x < 0 \\ 1 - 2x, & \text{if } 0 \le x \le 1 \\ x - 2, & \text{if } x > 1 \end{cases}$

5. Redo Exercise 4 with Piecewise.

6. Plot each of the functions:

    a) $f(x) = (x - \text{Floor}(x))^2$

    b) $f(x) = x(-1)^{\text{Floor}(x)}$

7. The function `Mod` provides an easy way to create a *periodic* function from a simpler function that describes a single period. For example, the function

    ```
    f0[x_] := √(1 - (x - 1)²)
    ```

    describes the top half of the circle with radius 1 centered at $(1, 0)$. Plot its graph by entering

    ```
    Plot[f0[x], {x, 0, 2},
      AspectRatio → Automatic, PlotRange → {-.5, 1.5}]
    ```

    The corresponding periodic function with period 2 is

    ```
    f[x_] := f0[Mod[x, 2]]
    ```

    In a similar manner, define and plot (for $0 \le x \le 9$) a periodic function with period 3 that agrees with $f_0(x) = e^{-x}$ on the interval $0 \le x < 3$.

8. The *unit step function* is defined by $u(x) = \begin{cases} 0, & x < 0, \\ 1, & x \ge 0. \end{cases}$ Its *Mathematica* name is `UnitStep`.

    a) Using `UnitStep`, plot the graph of u on $-1 \le x \le 2$.

    b) Plot $f(x) = u(x - 1)$ on $-1 \le x \le 2$ and express $f(x)$ in piecewise form.

    c) Plot $f(x) = u(x - 1)\sin(2\pi x)$ on $-1 \le x \le 2$ and express $f(x)$ in piecewise form.

    d) Plot $f(x) = u(x - 2) - h(x - 1)$ on $0 \le x \le 3$ and express $f(x)$ in piecewise form.

    e) Plot $f(x) = (u(x - 2) - u(x - 1))\sin(2\pi x)$ on $0 \le x \le 3$ and express $f(x)$ in piecewise form.

A.3 Equations

Certainly one of the most frequent mathematical tasks that we need to do is to solve an equation. In order to be able to solve equations with *Mathematica*, we first need to understand how equations are formed. The important thing to remember is that *double equal signs* are used to form an equation.

Actually, double equal signs constitute a *logical test* that returns either `True`, `False`, or the equation itself.

```
2 * 17 - 34 == 0
```

```
True
```

```
3 == 4
```

```
False
```

```
x² - 4 == 0
```

$-4 + x^2 == 0$

Notice, however, that *Mathematica* returns `True` only when the expressions on each side are identical. Only the most superficial simplification is done prior to the test, as in:

2 x + x - 2 == 3 x + 5 - 7

True

Notice that for this equation:

x^2 - 4 == (x + 2) (x - 2)

$-4 + x^2 == (-2 + x) (2 + x)$

a nontrivial operation must be done to one side or the other before the expressions become truly identical. Finally, notice that *Mathematica* returns an error message if we use a single equal sign improperly.

3 = 4

Set::setraw : Cannot assign to raw object 3. ≫

4

The single equal sign is used only for *assignments* such as

area = $\pi\, r^2$

$\pi\, r^2$

Mathematica has three functions for solving ordinary equations. These are `Solve`, `NSolve`, and `FindRoot`. `Solve` works very well on polynomial and many other algebraic equations.

Solve$\left[6\, x^3 - 23\, x^2 + 25\, x - 6 == 0,\ x\right]$

$\left\{\left\{x \to \frac{1}{3}\right\},\ \left\{x \to \frac{3}{2}\right\},\ \{x \to 2\}\right\}$

Solve$\left[x^4 - 2\, x^3 - x^2 + 6\, x - 6 == 0,\ x\right]$

$\left\{\{x \to 1 - i\},\ \{x \to 1 + i\},\ \left\{x \to -\sqrt{3}\right\},\ \left\{x \to \sqrt{3}\right\}\right\}$

Solve$\left[\sqrt{x - 1} + x == 4 + \sqrt{x + 4},\ x\right]$

$\{\{x \to 5\}\}$

`Solve` will also give solutions to trigonometric (or exponential/logarithmic) equations, but frequently gives a warning.

Solve$\left[2\, \text{Sin}[2\, x] == \text{Cos}[x]^2 - 1,\ x\right]$

Solve::ifun :

Inverse functions are being used by Solve, so some solutions may not be found;
use Reduce for complete solution information. ≫

$\left\{\{x \to 0\},\ \{x \to -\pi\},\ \{x \to \pi\},\ \left\{x \to \text{ArcCos}\left[-\frac{1}{\sqrt{17}}\right]\right\},\ \left\{x \to -\text{ArcCos}\left[\frac{1}{\sqrt{17}}\right]\right\}\right\}$

`Solve` will also find solutions of a system of equations. The equations must be given as elements of a list, *i.e.*, separated by commas and enclosed in braces.

$$\text{Solve}\left[\left\{x^2 - y == 1, -x + y == 1\right\}, \{x, y\}\right]$$

$\{\{y \to 0, x \to -1\}, \{y \to 3, x \to 2\}\}$

The solutions of polynomial equations of degree five or greater generally cannot be found in any exact form. Notice how *Mathematica* "avoids" the problem:

$$\text{Solve}\left[x^5 - 10 x^2 + 5 x + 1 == 0, x\right]$$

$\{\{x \to \text{Root}\left[1 + 5 \, \#1 - 10 \, \#1^2 + \#1^5 \, \&, \, 1\right]\},$
$\{x \to \text{Root}\left[1 + 5 \, \#1 - 10 \, \#1^2 + \#1^5 \, \&, \, 2\right]\},$
$\{x \to \text{Root}\left[1 + 5 \, \#1 - 10 \, \#1^2 + \#1^5 \, \&, \, 3\right]\},$
$\{x \to \text{Root}\left[1 + 5 \, \#1 - 10 \, \#1^2 + \#1^5 \, \&, \, 4\right]\},$
$\{x \to \text{Root}\left[1 + 5 \, \#1 - 10 \, \#1^2 + \#1^5 \, \&, \, 5\right]\}\}$

In such situations, we can always resort to numerical solutions. `NSolve` finds a numerical approximation to each solution of a polynomial equation, including complex solutions.

$$\text{NSolve}\left[x^5 - 10 x^2 + 5 x + 1 == 0, x\right]$$

$\{\{x \to -1.22065 - 1.89169 \, i\}, \{x \to -1.22065 + 1.89169 \, i\},$
$\{x \to -0.153102\}, \{x \to 0.66946\}, \{x \to 1.92494\}\}$

In many situations where `Solve` *is* successful, such as:

$$\text{Solve}\left[x^3 - 10 x^2 + 5 x + 1 == 0, x\right]$$

$\left\{\left\{x \to \dfrac{10}{3} + \dfrac{85}{3 \left(\frac{1}{2}\left(1523+9 \, i \, \sqrt{1691}\right)\right)^{1/3}} + \dfrac{1}{3}\left(\dfrac{1}{2}\left(1523 + 9 \, i \, \sqrt{1691}\right)\right)^{1/3}\right\},\right.$

$\left\{x \to \dfrac{10}{3} - \dfrac{1}{6}\left(1 + i \, \sqrt{3}\right)\left(\dfrac{1}{2}\left(1523 + 9 \, i \, \sqrt{1691}\right)\right)^{1/3} - \right.$

$\left.\dfrac{85\left(1 - i \, \sqrt{3}\right)}{3 \, 2^{2/3}\left(1523 + 9 \, i \, \sqrt{1691}\right)^{1/3}}\right\}, \left\{x \to \dfrac{10}{3} - \right.$

$\left.\left. \dfrac{1}{6}\left(1 - i \, \sqrt{3}\right)\left(\dfrac{1}{2}\left(1523 + 9 \, i \, \sqrt{1691}\right)\right)^{1/3} - \dfrac{85\left(1 + i \, \sqrt{3}\right)}{3 \, 2^{2/3}\left(1523 + 9 \, i \, \sqrt{1691}\right)^{1/3}}\right\}\right\}$

it may still be preferable to use `NSolve`:

$$\text{NSolve}\left[x^3 - 10 x^2 + 5 x + 1 == 0, x\right]$$

$\{\{x \to -0.152671\}, \{x \to 0.692369\}, \{x \to 9.4603\}\}$

Many equations *require* the use of `FindRoot`, which incorporates a numerical procedure. For example, consider

$$x^2 = \cos x.$$

`FindRoot` can find only one solution at a time and requires us to supply an initial guess at the solution we're looking for. An appropriate initial guess can usually be determined by examining a graph.

`Plot[{x², Cos[x]}, {x, 0, 1}]`

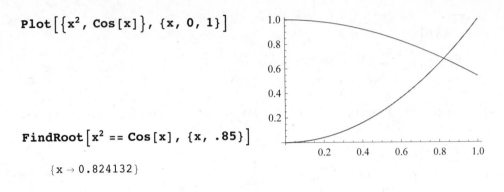

`FindRoot[x² == Cos[x], {x, .85}]`

   $\{x \to 0.824132\}$

◆ Exercises

1. The equation $x^3 = x + 1$ has one real solution. Find its exact value with `Solve` and its numerical value with `NSolve`.

2. Use `Solve` to find the solution(s) of each of the systems:

    a) $2x + 3y = 1$, $x^2 + y^2 = 1$ b) $3x - 2y = 5$, $7x + 3y = 2$

    c) $x + y + z = 2$, $x - y + z = 1$, $x^2 + y^2 + z^2 = 2$

3. Use `Solve` on the *underdetermined* system

    $$x + y + 2z = 2, \quad x - 2y + z = 1$$

    and interpret the result. Try each combination of *solve variables*: `{x,y,z}`, `{x,y}`, `{x,z}`, and `{y,z}`. Which gives the "cleanest" solution?

4. For each of the following functions, plot the graph to determine the approximate location of each of its zeros. Then find each of the zeros with `FindRoot`.

    a) $f(x) = e^{-x/2}$ b) $f(x) = x - 9\cos x$ c) $f(x) = x^2 - \tan^{-1} x$

5. For each of the following equations, plot both sides of the equation to determine the approximate location of each of its solutions in the specified interval. Then find each of the solutions with `FindRoot`.

    a) $\sin x \cos 2x = \cos x \sin 3x$, $0 \le x \le 2\pi$ b) $\sin x^2 = \sin^2 x$, $0 \le x \le \pi$

    c) $\tan x = x$, $0 \le x \le 3\pi$ d) $\sin x = e^{-x}$, $0 \le x \le 4$

6. Two spheres have a combined volume of 148 cubic inches and a combined surface area of 160 square inches. Find the radii of the two spheres.

7. Two spheres have a combined volume of 148 cubic inches and a combined surface area of 160 square inches. Find the radii of the two spheres.

8. An open-topped aquarium holds 40 cubic feet of water and is made of 60 square feet of glass. The length of the aquarium's base is twice its width. Find the dimensions of the aquarium.

9. a) Find the equation of the parabola that passes through the points $(-1, 1)$, $(1, 2)$, and $(2, 3)$.

    b) Find the cubic polynomial $f(x)$ such that $f(1) = f(2) = f(3) = 1$ and $f(4) = 7$.

A.4 Lists

Lists are ubiquitous in *Mathematica*. A list is anything that takes the form of a series of objects separated by commas and enclosed in braces, such as:

{a, b, c}

{{1, 3}, {2, 5}}

{x, 1, 2}

$\{x^2 + y == 2, \; 2x - y == 0\}$

Many built-in commands expect lists for certain arguments. For example, in

Plot[Cos[π Sin[x]], {x, 0, 2 π}]

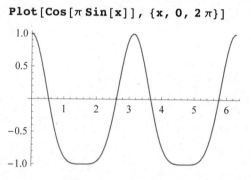

the second argument is a list that specifies the variable name and the interval over which to plot. In

Solve$\left[\{x^2 + y == 2, \; 2x - y == 0\}, \; \{x, y\}\right]$

$\{\{y \to 2\,(-1 - \sqrt{3}), \; x \to -1 - \sqrt{3}\}, \; \{y \to 2\,(-1 + \sqrt{3}), \; x \to -1 + \sqrt{3}\}\}$

each of the two arguments is a list, and the result is also a list (of lists).

■ Listable Functions

Most built-in *Mathematica* functions and operations are *listable*. When a listable function is applied to a list, it is applied to each element of the list and returns the result as a list. For example,

{1, 2, 3, 4, 5, 6, 7, 8, 9}^{2}

{1, 4, 9, 16, 25, 36, 49, 64, 81}

$\sqrt{\{\{\{2, 3\}, \{4, 5\}\}, \{\{6, 7\}, \{8, 9\}\}\}}$

$\{\{\{\sqrt{2}, \sqrt{3}\}, \{2, \sqrt{5}\}\}, \{\{\sqrt{6}, \sqrt{7}\}, \{2\sqrt{2}, 3\}\}\}$

$\dfrac{1}{\{1, 2, 3, 4, 5\}}$

$\left\{1, \dfrac{1}{2}, \dfrac{1}{3}, \dfrac{1}{4}, \dfrac{1}{5}\right\}$

```
{1, 2, 3} + {4, 5, 6}
```

```
{5, 7, 9}
```

```
{1, 2, 3} {4, 5, 6}
```

```
{4, 10, 18}
```

$2^{\{0,1,2,3,4,5,6,7,8,9,10\}}$

```
{1, 2, 4, 8, 16, 32, 64, 128, 256, 512, 1024}
```

```
Cos[{0, π / 4, π / 2, 3 π / 4, π}]
```

$$\left\{1, \frac{1}{\sqrt{2}}, 0, -\frac{1}{\sqrt{2}}, -1\right\}$$

■ The Parts of a List

Here's a simple list:

```
alist = {2, x, y, {a, b}}
```

```
{2, x, y, {a, b}}
```

Notice that it has four parts.

```
Length[alist]
```

4

Here's the third part:

```
alist[[3]]
```

y

The fourth part is itself a list:

```
alist[[4]]
```

```
{a, b}
```

Here's the second part of that sublist:

```
alist[[4, 2]]
```

b

This gives the first and third parts:

```
alist[[{1, 3}]]
```

```
{2, y}
```

This gives the first *through* the third parts:

```
alist[[1 ;; 3]]
```

```
{2, x, y}
```

This uses First to extract the first part:

```
First[alist]
```

2

This uses `Last` to extract the last part:

```
Last[alist]
```

```
{a, b}
```

■ Creating Lists

Mathematica provides three functions that are especially useful for creating lists. These are `Range`, `Table` and `NestList`.

■ Range

`Range` can be used with one, two, or three arguments. With one argument, it returns a list of consecutive natural numbers beginning with 1.

```
Range[10]
```

```
{1, 2, 3, 4, 5, 6, 7, 8, 9, 10}
```

`Range[a, b]` returns a list containing $a, a + 1, a + 2, ..., a + n$, where $a + n \le b < a + n + 1$.

```
Range[4.5, 15.1]
```

```
{4.5, 5.5, 6.5, 7.5, 8.5, 9.5, 10.5, 11.5, 12.5, 13.5, 14.5}
```

A third argument specifies the increment. (The default is 1.)

$$\texttt{Range}\left[0, 1, \frac{1}{10}\right]$$

$$\left\{0, \frac{1}{10}, \frac{1}{5}, \frac{3}{10}, \frac{2}{5}, \frac{1}{2}, \frac{3}{5}, \frac{7}{10}, \frac{4}{5}, \frac{9}{10}, 1\right\}$$

■ Table

`Table` provides an easy way of constructing many kinds of lists. The following computations illustrate its use.

$$\texttt{Table}\left[k^2, \{k, 10\}\right]$$

```
{1, 4, 9, 16, 25, 36, 49, 64, 81, 100}
```

```
Table[Table[i - j, {i, 4}], {j, 4}]
```

```
{{0, 1, 2, 3}, {-1, 0, 1, 2}, {-2, -1, 0, 1}, {-3, -2, -1, 0}}
```

```
% // MatrixForm
```

$$\begin{pmatrix} 0 & 1 & 2 & 3 \\ -1 & 0 & 1 & 2 \\ -2 & -1 & 0 & 1 \\ -3 & -2 & -1 & 0 \end{pmatrix}$$

▪ NestList

NestList creates lists whose elements are terms of a *recursive sequence*; that is, given a function f and a starting point a_1, it creates a list containing $a_1, a_2, a_3, ..., a_n$, where $a_{k+1} = f(a_k)$. For example, ten terms of the arithmetic sequence defined by

$$a_1 = 3, \ a_{k+1} = 2a_k - 1, \quad k = 1, 2, 3, ...$$

can be computed by first defining

    f[x_] := 2 x - 1

and then entering

    NestList[f, 3, 9]

    {3, 5, 9, 17, 33, 65, 129, 257, 513, 1025}

Notice that NestList has three arguments. The first is the name of the function, the second is the first member of the list, and the third is the number of "steps" to be computed (which is one less than the length of the resulting list).

▪ Manipulating Lists

▪ Flatten

There are occasions when we need to simplify lists by "merging" smaller lists that it contains. The Flatten command does this. For example:

    Flatten[{{1, 2, 3}, {4, 5}}]

    {1, 2, 3, 4, 5}

    Flatten[{{{x, y}, {3}}, {4, q}}]

    {x, y, 3, 4, q}

▪ Append and Prepend

We will often need to add elements to the end or beginning of an existing list. These tasks can be done with Append and Prepend. Here are two examples:

    Append[{1, 2}, 3]

    {1, 2, 3}

    Prepend[{1, 2}, 0]

    {0, 1, 2}

▪ Union and Join

Merging two or more lists into one can be done with Union or Join. Union does *not* maintain the order of elements:

    Union[{a, 2}, {v, 8}, {h, s}]

    $\left\{2, \ \dfrac{4324}{1209}, \ 8, \ h, \ s, \ v\right\}$

but `Join` does:

```
Join[{a, 2}, {v, 8}, {h, s}]
```

$$\left\{ \frac{4324}{1209}, 2, v, 8, h, s \right\}$$

■ Map and Apply

■ Map

A very useful method for applying a non-listable function to each element of a list is provided by `Map`. Suppose we have a list of ordered pairs of numbers such as

```
points = Table[{2 i, 5 - i}, {i, 8}]
```

```
{{2, 4}, {4, 3}, {6, 2}, {8, 1},
  {10, 0}, {12, -1}, {14, -2}, {16, -3}}
```

and we would like to create a list containing the sums of the numbers in each ordered pair in the list. To do this, we can create the function

```
addpairs[{x_, y_}] := x + y
```

and "map" it through the list of ordered pairs:

```
Map[addpairs, points]
```

```
{6, 7, 8, 9, 10, 11, 12, 13}
```

As an exercise, explain what goes on in the following:

```
Map[Flatten, {{3, {5, 6}}, {a, {b, c}}}]
```

$$\left\{ \{3, 5, 6\}, \left\{ \frac{4324}{1209}, \frac{99452}{1209}, \frac{34592}{20553} \right\} \right\}$$

■ Apply

Suppose that we have a function of two variables, say

```
vol[r_, h_] := π r² h
```

and that we would like to compute its value at a pair of numbers in a list, such as

```
measurements := {3.47, 5.12}
```

Now entering

```
vol[measurements]
```

```
vol[{3.47, 5.12}]
```

does not work. A very inconvenient, but effective, workaround is

```
vol[measurements[[1]], measurements[[2]]]
```

```
193.677
```

But a far simpler and more versatile approach is provided by `Apply` function:

```
Apply[vol, measurements]
```

```
193.677
```

It is usually easy to avoid such a situation (by defining `vol[{r_, h_}] := π r² h` in this case), but `Apply` does give us very nice way to compute the sum or product of the elements of a list:

```
Apply[Plus, {2, 5, 8, 12, 13}]
```

40

```
Apply[Times, {2, 5, 8, 12, 13}]
```

12 480

◆ Exercises

1. a) Use `Table` and `Prime` to generate a list of the first 100 prime numbers.

   b) Generate the same list using only `Prime` and `Range`.

2. Generate a list of values of the function $f(x) = \dfrac{\sin x}{x}$ for $x = .1, .2, ..., 1$, first using `Table`, then without `Table`.

3. Generate a list of the first 50 odd natural numbers, using:

   a) `Table`;   b) `Range`;   c) `NestList`

4. Generate a list of the first 21 powers of 2 (beginning with 2^0), using:

   a) `Table`;   b) `Range`;   c) `NestList`

5. Use `Table` to generate a list of ordered pairs $(x, f(x))$ for $x = 0, \dfrac{\pi}{12}, \dfrac{2\pi}{12} ..., \pi$, where $f(x) = \sin x$. Can you think of a way to do this without `Table`?

6. Create a list named `waves` that contains $\dfrac{1}{k} \sin kx$ for $k = 1, 2, 3, 4, 5$. Then plot the expressions in `waves` by entering

   ```
   Plot[Evaluate[waves], {x, 0, 2 π}]
   ```

   Now enter

   ```
   colors = Map[Hue, Range[.4, 1, .15]]
   Plot[Evaluate[waves], {x, 0, 2 π}, PlotStyle → colors]
   ```

   Then enter

   ```
   grays = Map[GrayLevel, Range[.8, 0, -.2]]
   Plot[Evaluate[waves], {x, 0, 2 π}, PlotStyle → grays]
   ```

7. a) The function

   ```
   f[x_] := x⁵ - 2 x² - 3 x + 3
   ```

   has three real zeros. Plot its graph, and create a list named `guesses` that contains a rough estimate of each of the zeros.

   b) Define the function

   ```
   getZero[guess_] := FindRoot[f[x], {x, guess}]
   ```

   and find all three zeros of f with one command by entering

   ```
   Map[getZero, guesses]
   ```

A.5 Rules

Understanding rules is essential to making efficient use of *Mathematica*. For example, note that the Solve command returns its results as lists of rules:

$$\text{soln} = \text{Solve}\left[\left\{x^2 + x + y^2 == 2, \; 2x - y == 1\right\}, \; \{x, y\}\right]$$

$$\left\{\left\{y \to \tfrac{1}{5}\left(-2 - \sqrt{29}\right), \; x \to \tfrac{1}{10}\left(3 - \sqrt{29}\right)\right\},\right.$$

$$\left.\left\{y \to \tfrac{1}{5}\left(-2 + \sqrt{29}\right), \; x \to \tfrac{1}{10}\left(3 + \sqrt{29}\right)\right\}\right\}$$

To convert this answer to a list of pairs of numbers, we apply the rules to the list {x,y} as follows:

$$\{x, y\} \; /. \; \text{soln}$$

$$\left\{\left\{\tfrac{1}{10}\left(3 - \sqrt{29}\right), \; \tfrac{1}{5}\left(-2 - \sqrt{29}\right)\right\}, \; \left\{\tfrac{1}{10}\left(3 + \sqrt{29}\right), \; \tfrac{1}{5}\left(-2 + \sqrt{29}\right)\right\}\right\}$$

Anticipating this in advance, we might have combined these steps by entering

$$\{x, y\} \; /. \; \text{Solve}\left[\left\{x^2 + x + y^2 == 2, \; 2x - y == 1\right\}, \; \{x, y\}\right]$$

$$\left\{\left\{\tfrac{1}{10}\left(3 - \sqrt{29}\right), \; \tfrac{1}{5}\left(-2 - \sqrt{29}\right)\right\}, \; \left\{\tfrac{1}{10}\left(3 + \sqrt{29}\right), \; \tfrac{1}{5}\left(-2 + \sqrt{29}\right)\right\}\right\}$$

The name of the *slash-dot* object "/." that we use to apply rules is ReplaceAll. The following are some simple examples that illustrate its use.

$$\sqrt{x^2 - x + 1} \; /. \; x \to 3$$

$$\sqrt{7}$$

$$x + y \; /. \; y \to x$$

$$2x$$

$$xy + yz + xz \; /. \; \{x \to a, \; y \to b + c, \; z \to 5\}$$

$$5a + 5(b + c) + a(b + c)$$

◆ Exercises

1. Trigonometric identities provide a good context in which to learn about rules and gain a bit of insight into symbolic computation in general. For example, the sine addition formula can be applied via the rule

    sinAdd := Sin[x_ + y_] → Sin[x] Cos[y] + Cos[x] Sin[y]

   Notice what happens when the rule is applied to $\sin(3x + 5y)$:

    Sin[3 x + 5 y] /. sinAdd

    Cos[5 y] Sin[3 x] + Cos[3 x] Sin[5 y]

   The same rule provides the sine difference formula as well.

    Sin[t - φ] /. sinAdd

    Cos[φ] Sin[t] − Cos[t] Sin[φ]

This rule also handles expressions with three or more summands, provided we use `//.` (`ReplaceRepeated`) instead of `/.`:

> **`Sin[a + b + c] //. sinAdd`**
>
> `Cos[b + c] Sin[a] + Cos[a] (Cos[c] Sin[b] + Cos[b] Sin[c])`

a) Enter the definition of `sinAdd` and construct a similar rule, `cosAdd`, for the cosine addition formula. Test both rules on several different expressions.

b) Enter the following multiple-angle expansion formula for sine:

> **`sinMult := Sin[n_Integer x_] → Sin[(n - 1) x + x] /. sinAdd`**

Check that this rule works properly by entering

> **`{Sin[2 x], Sin[3 x]} //. sinMult`**
>
> $\{2 \, Cos[x] \, Sin[x], \, 2 \, Cos[x]^2 \, Sin[x] + Cos[2 \, x] \, Sin[x]\}$

c) Construct a similar rule, `cosMult`, for the multiple-angle expansion formula for cosine. Test it on a few expressions.

d) Notice the result of repeatedly applying all four rules (followed by Expand) by entering

> **`Sin[2 x + y] //. {sinAdd, cosAdd, sinMult, cosMult} // Expand`**

Then enter `Simplify[%]` to verify that the expansion is correct.

e) Define the function

> **`trigExpand[expr_] :=`**
> **` expr //. {sinAdd, cosAdd, sinMult, cosMult} // Expand`**

and test it by entering

> **`Cos[x + 2 y] + Sin[3 x - y] // trigExpand`**
> **`% // Simplify`**

f) Finally, create a table of multiple angle formulas for sine by entering

> **`Table[{Sin[k x], Sin[k x] // trigExpand}, {k, 1, 5}] // TableForm`**

and create a similar table of multiple angle formulas for cosine.

2. a) Use `NSolve` to find the zeros of the polynomial $f(x) = x^5 - 4x^4 + 12x^2 - 9x + 1$. Convert the result to a list of numbers.

b) Compute the sum of the zeros of f using `Total`. Combine the entire process into a single command.

c) Repeat the process in parts (a) and (b) after changing the coefficient of x^4 to 3, and then once again after changing the coefficient of x^4 to 1. Try changing the other coefficients to see if they affect the result. What do you conjecture about the sum of the zeros of a fifth-degree polynomial?

d) Compute the *product* of the zeros of f using `Apply` and `Times`. (See section 1.4.) Experiment with the coefficients to determine which affect the result. What do you conjecture about the product of the zeros of a fifth-degree polynomial?

e) Experiment with a few polynomials of other degrees. Do your conjectures depend on degree? Also, are your conjectures consistent with *linear* polynomials?

A.6 Graphics

■ Graphics Objects and **Show**

Graphics commands such as `Plot` create and display **graphics objects**.

graph1 = Plot[Sin[x] Cos[10 x], {x, 0, 2 π}]

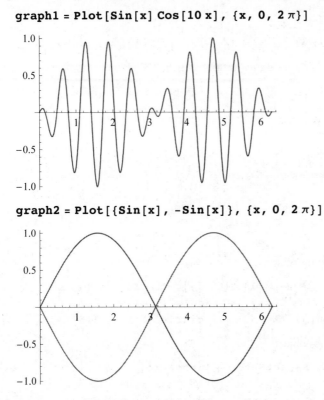

graph2 = Plot[{Sin[x], -Sin[x]}, {x, 0, 2 π}]

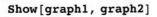

The `Show` command displays graphics objects, which may consist of two or more combined graphics objects.

Show[graph1, graph2]

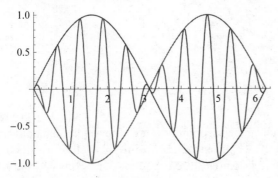

■ Graphics Primitives

Graphics primitives are the simple objects of which more complex graphics objects are built. Two- dimensional graphics primitives include `Point`, `Line`, `Circle`, `Disk`, `Rectangle`, `Polygon`, and `Text`.

The following defines a graphics primitive consisting of a series of line sergments connecting the specified points:

```
zigzag :=
 Line[{{1, 2}, {1, 1}, {3, 2}, {2, 2}, {1, 0}, {3, 1}, {2, 1}, {2, 0}}]
```

The `Graphics` command creates a graphics object from the graphics primitive.

```
Graphics[zigzag]
```

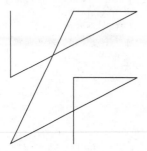

Here is a list of graphics primitives:

```
shapes = {Rectangle[{-2, 1}, {0, 2}], Circle[{1, 1}, 1],
    Disk[{0, 0}, .7], Text["rectangle", {-1.5, .8}],
    Text["disk", {-1, 0}], Text["circle", {1.5, 1.5}]};
```

This is the resulting graphics object:

```
Graphics[shapes, AspectRatio → Automatic]
```

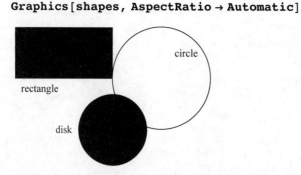

■ Graphics Directives

Graphics directives affect the way graphics primitives are displayed. Common graphics directives include `Opacity`, `PointSize`, and `Thickness`, as well as common colors such as `Red`, `Blue`, `Orange`, and so on. (`RGBColor`, `Hue`, `GrayLevel`, and `CMYK-Color` are available for mixing your own colors.)

A graphics directive is associated with a graphics primitive by creating a list of the form {*directive, primitive*}. More than one primitive can be specified by creating a list of the form {*directive1, directive2, ..., primitive*}. The following suggests the many possibilities:

```
redRect = {Red, Opacity[.67], Rectangle[{-1, 1}, {1, 2}]};
thickCircle =
  {Thickness[.02], Opacity[.5], Green, Circle[{1, 1}, 1]};
purpleDisk = {Purple, Opacity[.4], Disk[{0, 1}, .7]};
Graphics[{redRect, thickCircle, purpleDisk}]
```

Often graphics directives are provided through *options* such as PlotStyle, AxesStyle, and Background. For example,

$$\texttt{Plot}\big[\texttt{x}^2, \{\texttt{x, -1, 1}\}, \texttt{PlotStyle} \rightarrow \{\texttt{Thick, White}\},$$
$$\texttt{AxesStyle} \rightarrow \texttt{Green, Background} \rightarrow \texttt{GrayLevel}[.5]\big]$$

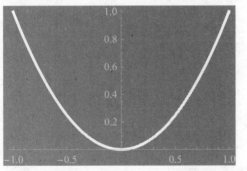

■ Suppressing and Combining Graphics

Suppose that we want to plot the parabola $y = x^2$ along with the circle of radius $1/2$ centered at $(0, 1/2)$. The following assigns names to plots of the parabola and the circle. *The output of each is surpressed by a semicolon.* (That's new in *Mathematica* 6.)

```
curve = Plot[x², {x, -1.4, 1.4}];
circ = Graphics[Circle[{0, .5}, .5]];
```

We can now combine the two graphics with `Show`.

`Show[curve, circ, AspectRatio → Automatic]`

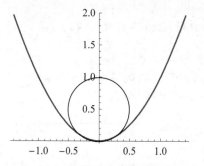

The same thing could be accomplished all at once by entering

$$\text{Show}\Big[\text{Plot}\big[\text{x}^2,\ \{\text{x},\ -1.4,\ 1.4\}\big],$$
$$\text{Graphics}[\text{Circle}[\{0,\ .5\},\ .5]],\ \text{AspectRatio} → \text{Automatic}\Big]$$

(This behavior is new in *Mathematica* 6; in earlier versions the preceding command would have produced three separate plots.)

■ `GraphicsRow` and `GraphicsGrid`

Two very useful graphics commands are `GraphicsRow` and `GraphicsGrid`. With these we can create composite graphics objects containing rectangular arrays of individual graphics objects. For instance, let's create the following four graphics objects:

```
segment = Graphics[Line[{{0, 0}, {2, 2}}]];
circ = Graphics[Circle[{0, 0}, 1]];
parabola = Plot[x², {x, -1, 1}, Axes → None];
rect = Graphics[{Gray, Rectangle[{0, 0}, {1, 1}]}];
```

The following shows these four graphics objects in a one-by-four array:

`GraphicsRow[{segment, circ, parabola, rect}]`

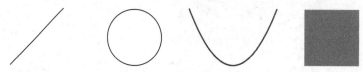

We would get a two-by-two array instead if we enter

`GraphicsGrid[{{segment, circ}, {parabola, rect}}]`

◆ Exercises

1. Predict the result of each of the following commands before entering it.

```
GraphicsRow[Table[Graphics[{Thickness[t], Orange, Circle[]}],
  {t, .02, .1, .02}]]
```

```
GraphicsRow[Table[
  Graphics[{PointSize[t], Point[{0, 0}]}], {t, .1, 1.1, .25}]]
```

```
GraphicsRow[Table[Graphics[{col, Disk[]}],
  {col, {Red, Blue, Green, Purple}}]]
```

```
GraphicsRow[Table[Graphics[{col, Disk[]}],
  {col, NestList[Lighter, Red, 3]}]]
```

```
GraphicsRow[
  Table[Graphics[{RGBColor[h, 0, 1 - h], Disk[]}], {h, 0, 1, .1}]]
```

```
Graphics[Table[{Hue[RandomReal[]],
  Disk[{Cos[t], Sin[t]}, .25]}, {t, π / 6, 2 π, π / 6}]]
```

A.7 Animate and Manipulate

One of the most instructive and fun features of *Mathematica* has always been its ability to animate graphics. New in *Mathematica* 6 is the function `Animate`, which provides a convenient, simple mechanism for creating animations.

For the sake of comparison, let's first create a simple table containing graphs of $y = \sin(x - \phi)$ for various values of the phase shift ϕ. Here ϕ will go from 0 to 2π in steps of 1 (by default).

```
Table[Plot[Sin[x - ϕ], {x, 0, 6 π},
  PlotRange → {{0, 6 π}, {-1.1, 1.1}}, AspectRatio → .2,
  Ticks → {Range[6] π, None}], {ϕ, 0, 2 π}]
```

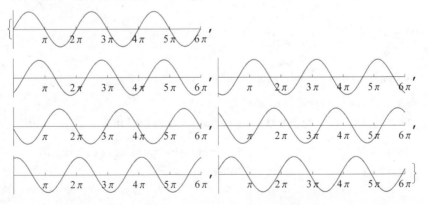

If we change `Table` to `Animate`, we get an animation instead, in which ϕ goes from 0 to 2π *continuously*. The controls allow you to start, stop, slow down, speed up, and reverse the animation. Moreover, the slider lets you "manually" move forward or backward through the animation.

```
Animate[Plot[Sin[x - ϕ], {x, 0, 6 π},
  PlotRange → {{0, 6 π}, {-1.1, 1.1}}, AspectRatio → .2,
  Ticks → {Range[6] π, None}], {ϕ, 0, 2 π}]
```

$$\phi \quad \begin{array}{cccccc} \pi & 2\pi & 3\pi & 4\pi & 5\pi & 6\pi \end{array}$$

Using `Manipulate` instead, we get only the slider control. However, clicking on the ⊞ icon at the right end of the slider will reveal animation controls.

```
Manipulate[
  Plot[Sin[x - ϕ], {x, 0, 6 π}, PlotRange → {{0, 6 π}, {-1.1, 1.1}},
  AspectRatio → .2, Ticks → {Range[6] π, None}], {ϕ, 0, 2 π}]
```

☐ Notice in the preceding examples the specification of the `PlotRange` option. This is generally necessary to ensure that each plot created corresponds to the same rectangle in the plane, thereby producing an animation in which any fixed point remains still.

`Animate` and `Manipulate` can be used to animate or manipulate any type of expression, not just graphics.

$$\text{Manipulate}\big[\text{Expand}\big[(a + b)^n\big], \{n, 1, 20, 1\}\big]$$

n

$$a^7 + 7 a^6 b + 21 a^5 b^2 + 35 a^4 b^3 + 35 a^3 b^4 + 21 a^2 b^5 + 7 a b^6 + b^7$$

Here are two simple examples that use a different `ControlType`:

$$\text{Manipulate}\big[\text{Expand}\big[(a + b)^n\big],$$
$$\{n, 1, 15, 1\}, \text{ControlType} \to \text{SetterBar}\big]$$

$$\text{Manipulate}\big[x^y, \{x, 1, 15, 1, \text{ControlType} \to \text{SetterBar}\},$$
$$\{y, 1, 15, 1, \text{ControlType} \to \text{SetterBar}\}\big]$$

◆ Exercises

1. The following creates a `Manipulate` object that plots the graph of $y = \sin(ax) + \sin(bx)$ for manipulable *angular frequencies* a and b of the individual terms. Enter the command and experiment with the two sliders. Then describe what you observe whenever the value of a "crosses over" the value of b or vice versa.

```
Manipulate[Plot[Sin[a t] + Sin[b t],
    {t, 0, 5}, PlotRange → {-2, 2}, PlotPoints → 40],
    {a, 50, 100}, {{b, 100}, 50, 100}]
```

This demonstrates the same phenomenon with sound:

```
Play[Sin[(950 + 10 t) t] + Sin[1000 t], {t, 0, 5}]
```

And this is the same thing in stereo:

```
Play[{Sin[(950 + 10 t) t], Sin[1000 t]}, {t, 0, 5}]
```

This is the phenomenon known as *beats*, which is useful in tuning string instruments by ear.

2. In this exercise, we will build up some simple and strangely interesting pieces of *Mathematica*-generated "art." Begin by entering

```
r := RandomReal[];
Show[Graphics[Line[{{r, r}, {r, r}}]], PlotRange → {{0, 1}, {0, 1}}]
```

This simply plots a random line segment within the square $-1 \le x \le 1, -1 \le y \le 1$. (Re-enter this a couple of times to observe the difference in the results.) Now enter the following several times. You should observe forty random segments each time.

```
Graphics[Table[Line[{{r, r}, {r, r}}], {40}],
    PlotRange → {{0, 1}, {0, 1}}]
```

Let's now give random color and thickness to the segments. Enter this a few times:

```
Graphics[Table[
    {RGBColor[r, r, r], Opacity[e^-1.5 r], Thickness[.01 + .02 r],
        Line[{{r, r}, {r, r}}]}, {40}], PlotRange → {{0, 1}, {0, 1}}]
```

Now create a composite graphic by entering

```
GraphicsRow[Table[Graphics[Table[{RGBColor[r, r, r],
    Thickness[.003 + .01 r], Line[{{r, r}, {r, r}}]}, {40}]], {5}]]
```

Enter the following to produce an animated piece on a black background:

```
Animate[Graphics[k; Table[
    {RGBColor[r, r, r], Opacity[.9 e^-2 r], Thickness[.01 + .02 r],
        Polygon[{{r, r}, {r, r}, {r, r}}]}, {30}],
    PlotRange → {{0, 1}, {0, 1}}, Background → Black], {k, 1, 20}]
```

Repeat the animation above, replacing the `Line[{{r,r},{r,r}}]` primitive with

a) `Circle[.25{1+2r,1+2r},7 r/5]`

b) `Disk[{r,r}, .5r]`   c) `Polygon[{{r,r},{r,r},{r,r}}]`

A.8 Avoiding and Getting Out of Trouble

❀ A Top Eleven List: Causes of *Mathematica* Problems

11. Forgetting that the natural log function is Log, not Ln

This is not peculiar to *Mathematica*; many advanced texts use this convention. However, if this bothers you, you can always define Ln[x_]:=Log[x].

10. Typing an equation with one equal sign (=) instead of two (==)

A single equal sign is used only for assignments; an equation requires two. Recovering from this mistake often requires that you Clear a variable.

9. Forgetting to type a space between multiplied expressions

For example, if you accidentally type xSin[x], *Mathematica* assumes that you are refering to a function named xSin.

8. Using parentheses instead of brackets or braces (or vice-versa)

Parentheses, brackets, and braces have very specific and different uses. Parentheses are used only for grouping within expressions, brackets enclose function arguments, and braces enclose members of a list.

7. Forgetting to load a package before referencing something in it

To correct the "shadowing" problem that results from this, you can enter Remove[*object*] where *object* is what you tried to use prior to loading the package.

6. Entering a command that relies on a previous definition that has not been entered during the current session

Whenever you resume work from a previous session, be sure that you re-enter commands in order from the top of your *Mathematica* notebook.

5. Doing an enormous symbolic computation instead of a simple numerical computation

See ▪ Symbolic versus Numerical Computation below.

4. Making multiple definitions for one variable or function name

See ▪ Multiple Definitions and Using the Question Mark below.

3. Spelling errors (including capitalization)

Enough said.

2. Forgetting to use a Blank when defining a function

See Section A.3.

1. Forgetting to save your work before the inevitable crash

A word to the wise...

■ What to Do When You Run into Trouble

□ *Check for spelling mistakes, typos, and other syntax errors.*

□ *Look for online help.*

You can access the Documentation Center through the Help menu or by simply pressing your "help" key. If you want help on a particular command, option, etc., highlight that item before pressing "help."

□ *Clear variable names.*

Remember that entering Clear[*var1*, *var2*,...] clears variables.

□ *Clear everything.*

Here's a quick way to clear *all* previous definitions:

```
ClearAll["Global`*"]
```

□ *Quit and restart the kernel.*

Do this by choosing Quit Kernel: Local from your Kernel menu. You can then choose Start Kernel: Local from your Kernel menu or simply enter a command to start a new kernel session.

□ *Quit and restart Mathematica.*

Be sure to save your work first.

□ *Quit Mathematica and restart your computer.*

Again, be sure to save your work.

□ *Quit and restart your day. (Just kidding.)*

Seriously though, if you get frustrated, *take a break!*

▯ **Important note:** Although *Mathematica* remembers everything you enter during a particular session, it does not remember anything from a previous session or anything prior to clearing all variables or restarting the kernel. Since much of what you do in *Mathematica* depends on previously entered commands, you must be careful to reënter the commands that are needed after clearing all variables or restarting the kernel.

■ Interrupting Calculations

You will occasionally enter a command that takes *Mathematica* a very long time to evaluate. To stop a computation, select Abort Evaluation from the Evaluation menu. The keyboard shortcut for this is ⌘-. (command-period) on a Macintosh and CTRL-C on an Windows PC.

It is often necessary to press these keys repeated to interrupt a calculation, and sometimes there is no alternative but to quit the kernel.

■ Interpreting *Mathematica* Output When Things Don't Work

In many circumstances, *Mathematica* will give you a useful error message when a bad command is entered. Here are two examples:

Plot$\left[x^2, \{0, 1\}\right]$

Plot::pllim : Range specification {0, 1} is not of the form {x, xmin, xmax}. ≫

Plot$\left[x^2, \{0, 1\}\right]$

Solve$\left[x^2 + 5\,x = 2, x\right]$

Set::write : Tag Plus in $5\,x + x^2$ is Protected. ≫
Solve::eqf : 2 is not a well-formed equation. ≫

Solve[2, x]

However, it is very common for *Mathematica* simply to give a problematic command back to you with no message. *Mathematica* does this whenever the syntax is correct, but no currently defined rules affect the result. For example,

aFunctionNotEntered[0]

aFunctionNotEntered[0]

and

Ln[1]

Ln[1]

▯ When *Mathematica* simply gives a command back to you with no error message, it means that the syntax is okay, but something in the command is unrecognizable.

■ Symbolic Versus Numerical Computation

It is easy to run into major trouble by inadvertently asking *Mathematica* to create a huge symbolic expression. This is most likely to happen as a result of doing some kind of recursive calculation. For example, suppose we want to calculate several terms in the sequence defined by

$$x_0 = 1 \text{ and } x_{k+1} = 3 \sin x_k - x_k \text{ for } k = 0, 1, 2, \ldots$$

Here is a typical *Mathematica* approach: We'll define the function

f[x_] := 3 Sin[x] - x

and use NestList to compute terms in the sequence. This computes the first five terms:

NestList[f, 1, 4]

{1, -1 + 3 Sin[1], 1 - 3 Sin[1] - 3 Sin[1 - 3 Sin[1]], -1 + 3 Sin[1] +
   3 Sin[1 - 3 Sin[1]] + 3 Sin[1 - 3 Sin[1] - 3 Sin[1 - 3 Sin[1]]],
   1 - 3 Sin[1] - 3 Sin[1 - 3 Sin[1]] - 3 Sin[1 - 3 Sin[1] - 3 Sin[1 - 3 Sin[1]]] -
   3 Sin[1 - 3 Sin[1] - 3 Sin[1 - 3 Sin[1]] -
      3 Sin[1 - 3 Sin[1] - 3 Sin[1 - 3 Sin[1]]]]}

This is not exactly what we had in mind, is it? If we had asked for ten terms instead of five, the result would have filled several pages. (Try it.) If we had asked for thirty terms, and *if* the computation had *eventually* succeeded, the result would have contained more than 1/2 *billion* copies of the expression Sin! (Do yourself a favor; *don't* try it.)

So what should we do? We simply need to coerce *Mathematica* into doing the calculation numerically instead of symbolically, which is what we wanted to begin with! One simple way to do this is to start the sequence with the real number 1. instead of the integer 1. (Can you think of two other ways to accomplish the same thing?)

```
f[x_] := 3 Sin[x] - x; NestList[f, 1., 30]
```

{1., 1.52441, 1.47236, 1.51312, 1.48189, 1.50626,
  1.4875, 1.5021, 1.49082, 1.49959, 1.49281, 1.49807, 1.494,
  1.49716, 1.49471, 1.49661, 1.49514, 1.49628, 1.4954,
  1.49608, 1.49555, 1.49596, 1.49564, 1.49589, 1.4957,
  1.49585, 1.49573, 1.49582, 1.49575, 1.49581, 1.49576}

☐ *A useful tip*: When attempting a complicated computation, *start small!* In other words, see what happens when you do three steps before you try to do thirty.

■ Multiple Definitions and Using the Question Mark

Suppose that we enter

```
f[x] = x² + 3 x
```

$3 x + x^2$

and we then realize that we forgot the Blank that we need to put beside the variable. So we then enter

```
f[x_] = x² + 3 x
```

$3 x + x^2$

and everything seems fine. *Later*... when working on a different problem, we redefine f as

```
f[x_] = x - 2
```

$-2 + x$

This function behaves as we expect; we find that its graph is the expected straight line, etc. But then we enter

```
g[x_] = f[x]²
```

$\left(3 x + x^2\right)^2$

which does not give the function g that we expect. So what's going on here? *Mathematica* remembers our original, "erroneous" definition of the expression f[x].

♡ Using the Question Mark

To get information on any variable or other object, just type its name after a question mark. For example, to get information on f we'll enter

```
? f
```

Global`f

$f[x] = 3x + x^2$

$f[x\_List] := 3x$

$f[x\_] := Which[x \le -1, 1, -1 < x \le 1, -x, x > 1, -1]$

$f[x\_, y\_] := x + y$

This shows us that multiple definitions are associated with f. In fact, we could cause *Mathematica* to associate numerous other definitions with f:

```
f[x_, y_] := x + y
f[x_List] := 3 x
```

Now let's get information on f:

```
? f
```

Global`f

$f[x] = 3x + x^2$

$f[x\_List] := 3x$

$f[x\_] := Which[x \le -1, 1, -1 < x \le 1, -x, x > 1, -1]$

$f[x\_, y\_] := x + y$

When *Mathematica* encounters an expression involving f, it looks through the definitions associated with f until one makes sense for that expression. For example:

```
f[3, 5]
```

8

```
f[13]
```

11

```
f[{4, 7}]
```

{12, 21}

```
f[x]
```

$3x + x^2$

As you may well imagine, this behavior of *Mathematica* can potentially be the source of all kinds of trouble.

▢ The key to resolving difficulties caused by multiple definitions is to Clear the culprit variable. If you get into a really complicated mess, try quitting the kernel or entering

```
ClearAll["Global`*"]
```

The question mark is also useful for getting the "usage message" for built-in objects. Here are a few examples:

? Plot

Plot[f, {x, x_{min}, x_{max}}] generates a plot of f as a function of x from x_{min} to x_{max}.
Plot[{f_1, f_2, ...}, {x, x_{min}, x_{max}}] plots several functions f_i. ≫

? $DisplayFunction

$DisplayFunction gives the default setting
    for the option DisplayFunction in graphics functions. ≫

? NestList

NestList[f, *expr*, n] gives a list of the results of applying f to *expr* 0 through n times.
    ≫

? /.

expr /. *rules* applies a rule or list of rules in an
    attempt to transform each subpart of an expression *expr*. ≫

■ Memory

Some of the most common difficulties that arise when using *Mathematica* are memory related—or rather, *lack-of*-memory related. *Mathematica* consists of two applications—the **kernel** and the **front end**—working together. Each of these has its own memory.

■ Kernel Memory

The kernel is the part of *Mathematica* that does the computation. Many of the computations done by *Mathematica* involve highly complex algorithms and require a great deal of memory. In addition, the kernel remembers (by default) every command entered and every computation done in a given session. So it is easy to understand why running out of kernel memory—or experiencing poor performance due to use of virtual memory—can happen.

There are a couple of simple things that you can do to conserve memory:

☐ Set $HistoryLength to some small value such as 10 (*i.e.*, enter $History-Length=10). This causes the kernel to forget older input and output lines. The default value of $HistoryLength is Infinity.

☐ Use the Share command occasionally:

Share[]

1 010 880

This causes stored expressions to "share" subexpressions, thus reducing the amount of memory used. The output shows the number of freed bytes.

Also, see ■ Symbolic versus Numerical Computation in the preceeding section.

■ Front End Memory and File Size

Front-end memory and notebook file size usually only become an issue when your notebook contains a lot of graphics. While the computations that create a graphic are done by the kernel, the code that actually produces the graphic is stored in the front end's memory.

By deleting graphics cells—especially cells containing three-dimensional graphics or graphics created with a high value for the `PlotPoints` option—you can greatly decrease the amount of front-end memory used.

When you save your work, it is information in the front end's memory that you are saving—in the form of a *Mathematica* "notebook." When a notebook becomes very large, you can usually remedy the situation by deleting graphics cells before saving. Graphics can always be reproduced, as long as the commands are saved.

☐ A handy feature is the Delete All Output item in the Cell menu. This will let you quickly save the essence of your work in a very small file.

■ Exercises

1. Purposely commit each of the errors described in the eleven causes of problems outlined above—with the exception of numbers 4 and 1. In cases where no consequence is immediately evident, construct a subsequent scenario that exposes the error.

A.9 Turning a Notebook into a Report

You will likely be asked to put the work you do in *Mathematica* into a form that will be presentable enough to submit to your professor. Fortunately, the *Mathematica* front end serves as a very versatile word processor. In fact, this entire manual was written with *Mathematica*.

■ Cell Styles

You should always provide comments and narrative along with your calculations (whether you're using *Mathematica* or pencil and calculator). Any cell that contains text should be given Text Style by selecting Text from the Style submenu of the Format menu before you begin typing text into the cell. You can also give an existing cell Text Style by highlighting the cell bracket and selecting Text from the same menu.

You should also use Title, Section, Subsection cells, etc., to organize your notebook. These items are also in the Style submenu of the Format menu.

Aside from resulting in much nicer looking work, this is also important because when an Input cell contains text, all kinds of errors and garbage can result if it is accidentally evaluated.

> **For example notice what happens when I enter this**
>
> i enter example For happens notice this what when

or when I enter this.

Syntax::tsntxi : "this." is incomplete; more input is needed.

Syntax::tsntxi : "this." is incomplete; more input is needed.

Syntax::sntxi : Incomplete expression; more input is needed.

Moreover, if Input cells only contain valid *Mathematica* commands, it is possible to evalute all of them in order by selecting Evaluate Notebook from the Evaluation menu without making a big mess of your the notebook.

■ Page Breaks

Bad page breaks usually involve a large amount of blank space at the bottom of a page. Frequently this is caused by a graphic being just a little too large to fit on the current page. The default size of *Mathematica* graphics is larger than it usually needs to be for printing. So by resizing (*i.e.*, shrinking) graphics, you can avoid a lot of bad page breaks. A graphic can be resized by clicking on it and dragging a corner. Also, a good way to get smaller, consistently-sized graphics from Plot, for instance, is to use SetOptions:

```
SetOptions[Plot, ImageSize → 200]
```

How can you tell where page breaks will occur before you print? You can select Show Page Breaks from the Printing Settings submenu of the File menu. (In *Mathematica* 5.x, this is located in the Format menu.)

You can also *force* a page break between two cells by selecting Page Break from the Insert Menu.

■ Other Tips

In the Printing Settings submenu of the File menu, you'll see Printing Options... In the resulting dialog box, you can set margins and specify whether or not to print cell brackets.

In the Style Sheet submenu of the Format menu, you can choose from among several standard style sheets. The choice of style sheet affects the appearance of title and section cells, background color, etc. Experiment to find one that you like. But don't be surprised if you end up preferring the default.

◆ Exercises

1. Write a short but detailed report (2-3 pages) in *Mathematica* on any one of the following topics. Your report must include input, output, text, section, and title cells.

   a) The Rational Root Theorem for polynomials.

   b) How to find the inverse of a one-to-one function.

   c) Even functions and odd functions.

   d) The unit circle and the graphs of $\sin x$ and $\cos x$.

   e) The compound interest formula.

   f) Rational functions with slant asymptotes.

A.10 Miscellaneous Advice

■ The Cube Root Function

When you ask *Mathematica* for the cube root (or any odd root) of a negative number, it returns a complex number. This complex number is indeed the *principal value* the cube root function defined on the complex numbers.

$$\sqrt[3]{-8.}$$

`1. + 1.73205 i`

However, this is not what we want when we talk about the cube root function defined on the real numbers. A simple remedy is to define your own cube root function as follows:

`Cbrt[x_] := Sign[x] ` $\sqrt[3]{\text{Abs}[x]}$

`Plot[Cbrt[x], {x, -4, 4}]`

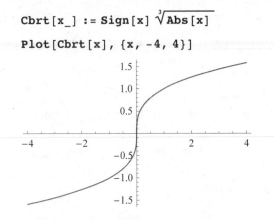

■ Custom Initialization

You may find that there are certain commands you want to enter every time you run *Mathematica*. For instance, if you prefer that the curves drawn by `Plot` are always thicker than the default thickness, you can avoid specifying that with the `PlotStyle` option every time you use plot by entering

`SetOptions[Plot, PlotStyle → Thickness[.005]]`

But that will only be in effect during the current kernel session; that is, the next time you start *Mathematica*, you'll have to enter that command again.

Assuming that you have appropriate priviledges on the computer on which you're working, you can execute a group of commands automatically each time the kernel starts up by placing those commands in a file named init.m that's located in a particular directory. An easy way to find and open that file is to enter

`ToFileName[{$UserBaseDirectory, "Kernel"}, "init.m"]`
`NotebookOpen[%]`

You can type in the commands you want executed automatically upon startup and then save and close the file. Actually, it's best to type the commands in an open notebook, make sure they work correctly, and then copy and paste them into the init.m file.

Here are a few examples of things you might want to put in init.m:

• This sets options for `Plot` so that it draws thicker curves and makes plots 3 inches wide.

```
SetOptions[Plot,
    {PlotStyle → Thickness[.005], ImageSize → 3 * 72}];
```

• This defines a real cube root function.

$$\text{Cbrt}[x\_] := \text{Sign}[x] \sqrt[3]{\text{Abs}[x]} ;$$

• This sets $HistoryLength to conserve memory.

```
$HistoryLength = 10
```

Index